SOLAR SYSTEM

Peter Ryan

Illustrated by Ludek Pesek
Diagrams by QED

The Viking Press · New York

SOLAR SYSTEM

Published in 1979 by The Viking Press
625 Madison Avenue, New York, N.Y. 10022

Library of Congress Cataloging in Publication Data
Ryan, Peter, 1939–
  Solar system.

  Includes index.
  1. Solar system.   I.   Title.
QB501.2.R89   523.2   78-15382
ISBN 0-670-65636-4

Printed in Great Britain

Set in Monophoto Plantin

DEDICATED TO PH

A Note on Units

Throughout this book, the term *billion* has been used, in the American sense, to denote one *thousand million* or 10^9 (the figure 1 followed by nine zeros). Thus *trillion* refers to 10^{12}, *quadrillion* to 10^{15}, *quintillion* to 10^{18} and *septillion* (a *trillion trillion*) to 10^{24}.

Measurements of distances, diameters and mass are given in metric units. A *kilometre* is about five eighths of a *mile*, a *metre* is about three inches more than a *yard* and there are about two and a half *centimetres* to an inch. A *kilogram* is almost exactly 2·2 *pounds*. A metric *tonne* (1,000 kilograms) is 204·6 pounds more than an American *short ton* and 35·4 pounds less than a British *long ton*.

One *Astronomical Unit* (AU), the average earth–sun distance, is 149·6 million kilometres. One *light year*, the distance travelled by light in one earth year, is 9·46 trillion kilometres or 63,240 AU.

In some places temperatures are given in *degrees Kelvin* (°K). This scale starts at 'absolute zero' which is almost exactly minus 273° on the *Centigrade* (C) scale. Thus 0 °C is almost exactly 273 °K.

CONTENTS

FOREWORD BY CARL SAGAN

Humanity is at this moment engaged in a remarkable enterprise – the first detailed reconnaissance of the solar system which we inhabit. No longer are the planets slow-moving and mysterious lights in the night sky. Instead they are becoming perceived as worlds. Hardy and intrepid unmanned spacecraft have obtained close-up views of the planets from Mercury to Jupiter. Space vehicles have landed on the moon, Venus, and Mars. Shortly we will have close-up images of Saturn as well. Some of these planets are small, rocky, and metallic like the earth; others are giant, cloudy balls of hydrogen like Jupiter. Not a one is closely similar to the earth. And the solar system is filled with innumerable smaller objects – moons, some with atmospheres, asteroids, comets, the particles which make up the rings of Saturn and Uranus. These other worlds are important not only in their own right, not only for the spirit of adventure and the zest for exploration which they arouse; but also because they are comparisons for the earth, indications of how things might have gone on this planet had initial conditions been somewhat different, or how things might go on the earth in the future if we are not careful. The exploration of the planets has both a romantic and a practical appeal.

Ages of exploration need their travel books, their descriptions of exotic and far-off lands. *Solar System* is a very good such guidebook for the beginning space traveller. Peter Ryan's text is straightforward, generally easy to understand, and an accurate reflection of the present state of knowledge. The many photographs – particularly those taken by unmanned planetary spacecraft – give a sense of reality to places which once were visited only in dreams. And Ludek Pesek's paintings provide an immediacy, a sense of place of these other worlds which will be stirring to explorers of all ages. Interest in the solar system is, of course, not restricted to the young; but the young have a special interest in it because it is they who will live in the future, in an age when there may be fast and frequent voyages between the earth and outposts of humanity on some other and distant worlds. For a glimpse of at least part of the future, read this book.

CARL SAGAN
Laboratory for Planetary Studies
Cornell University

May 1978

SOLAR SYSTEM

In the beginning there was a vast and black void. In the middle of all this absence lurked a single titanic blob, a mass of a septillion suns, the remains perhaps of an older universe. Its explosion ten to twenty billion years ago marks the beginning of our time. Billowing behind the shock waves came a gigantic cloud of gas, the primeval matter from which in time a hundred billion galaxies were born. One of these was ours, the Milky Way galaxy – today a lens-shaped flock of stars one hundred thousand light years across.

Scooped out of the primordial cloud by eddies in the cosmic current, the Milky Way began its life as a rolling spiral of gas. Condensed by gravity, its rate of spin – like that of a pirouetting ballerina – increased as its limbs were drawn more closely about its centre. The spinning promoted a gradual flattening and, as a first sprinkling of stars was formed, the gas cloud filled with light. Within the more massive of this first generation of stars nuclei of hydrogen and helium were fused to create nuclei of heavier elements. But these massive stars were unstable and short-lived. Their explosive deaths created brilliant supernovae and vast clouds of dust.

That, according to the 'big-bang' theory, is how it all began. Currently much less popular with astronomers is the rival 'steady-state' theory. By this account, the universe is in an eternal state of continuous creation. The stellar population – some trillion trillion stars, collected in a hundred billion galaxies – is more or less constant. When an old star fades, its place is taken by a new one, formed from freshly created matter.

Big bang or not, the galaxy was at least a few billion years old when a cloud of dust and gas, buffeted by the explosive birth, or perhaps death, of a nearby massive star, collapsed through interstellar space. Within a spiral arm of the Milky Way, two thirds of the way out from its centre, the nucleus of this rolling cloud of dust and gas became dense and hot, ignited and became a star, our sun. According to one theory, the remaining particles were then gathered into rings around the newborn sun. At some later stage, around 4·6 billion years ago, these particles were drawn together, ring by ring, to make up nine planets, at least thirty-three moons, thousands of asteroids, and billions of meteoroids and comets.

A sequence of three paintings depicts the ancestry and birth of our solar system. On the left, the first shows what the birth of our universe might have looked like, seen from afar.

A second painting, overleaf, shows the birth of the Milky Way galaxy (foreground) when a streamer of matter from the expanding 'big bang' gas cloud was set spinning by an eddy to become a galactic spiral. In this illustration the young galaxy is illuminated by the light of massive stars and exploding supernovae. Between them, enormous clouds of dust and gas contain the ingredients of sun-like stars, planets, asteroids, moons, meteoroids and comets.

On the next page, the third painting in the series records the creation of our solar system when, some five billion years ago, the sun's nuclear furnace was ignited at the dense centre of a rotating cloud of dust and gas.

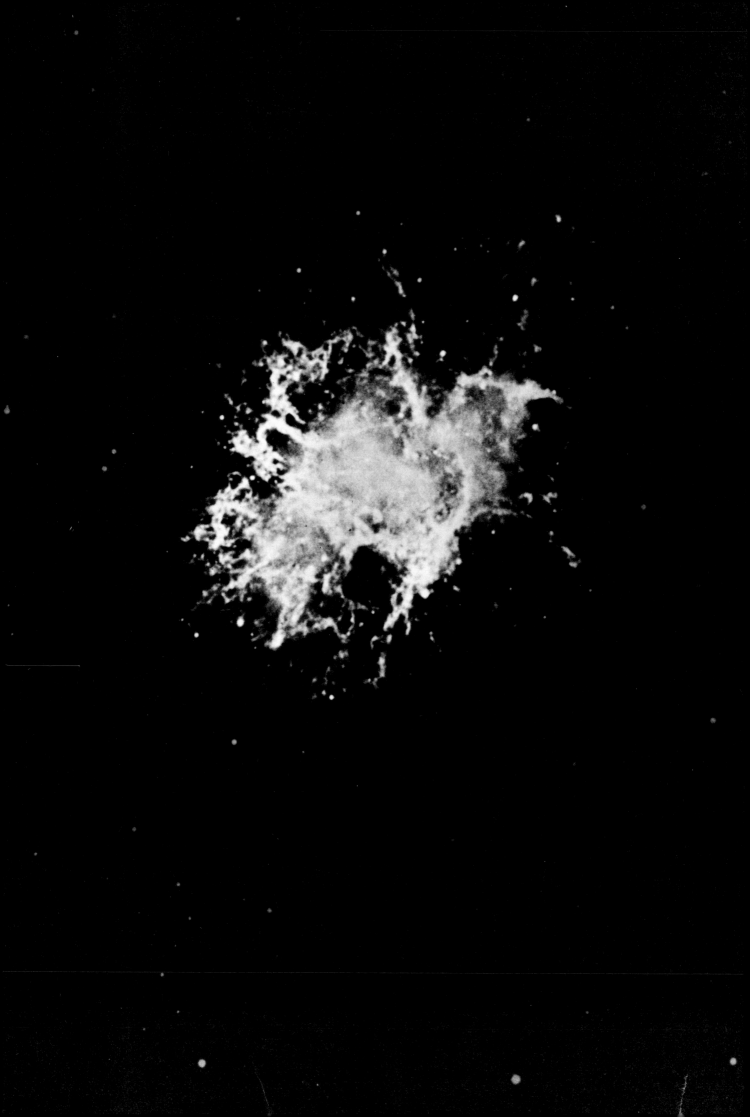

The innermost of the nine planets, Mercury, is similar to and scarcely larger than the earth's moon. This cratered globe has an elliptical orbit; its distance from the sun varies between three tenths and a little under five tenths of an astronomical unit – one astronomical unit, or AU, being the average earth–sun distance, 149,600,000 kilometres. The second planet, Venus, has a near-circular orbit. At any given moment its distance from the sun is always just over seven tenths of an AU. A barren rocky world, its surface is permanently hidden beneath thick clouds of sulphuric acid. About the same size as Venus, planet three has a more transparent atmosphere, a mixture of nitrogen and oxygen. The most noticeable surface feature is extensive flooding – nearly three quarters of the earth is under water. Our distance from the sun varies between 147,100,000 and 152,100,000 kilometres. Our single moon is the solar system's sixth largest. The fourth planet, Mars, is bigger than Mercury but smaller than the earth. A spherical red desert with polar ice caps, it has two tiny pock-marked moons, Phobos and Deimos. Mars, whose distance from the sun varies between about 1·4 and 1·7 AU, is the outermost member of the group called the 'inner planets'.

Except for the earth, and perhaps Venus, the inner planets and their moons are covered with craters. Formed by collisions during the final stages of accumulation, they show that bodies with diameters of up to 100 kilometres were once common in this part of the solar system. On earth, the possible melting of the entire planet some 4·6 billion years ago and the subsequent

---

Little bang. This gigantic cloud of dust and gas was produced by the explosion of a massive Milky Way star. The event – a supernova – was recorded by Chinese astronomers in 1054. Now called the Crab Nebula, its distance from the sun is around 6,300 light years.

As well as filling interstellar space with hydrogen and helium gas enriched by the nuclei of heavier elements, the explosion released an enormous amount of energy. Around five billion years ago, shock waves from a nearby supernova may have helped another vast cloud of dust and gas – already enriched by previous explosions – to become our solar system.

deformation of the land masses by continental movements and erosion have long since erased all but a few of these cosmic birthmarks. Today the traffic of asteroids, meteoroids and comets through inner-planet orbits is much reduced but it has not stopped completely. In 1908 a colliding comet did extensive damage to a Siberian forest and, as recently as 1972, a 1,000-tonne meteoroid bounced off the atmosphere over the state of Montana. Thought by some to be the would-be ingredients of a missing planet, the great majority of these flying rocks are concentrated in a series of bands beyond the orbit of Mars between around 2·2 and 3·5 AU. The largest of several thousand bodies, the asteroid Ceres has a diameter of about 1,000 kilometres.

Beyond these asteroids, by far the most massive planet, fifth from the sun, is the immense cloud-banded world of Jupiter. Orbiting the sun at a distance of between 5·0 and 5·5 AU, its diameter is more than eleven times that of the earth. The first of four giants, it has a family of at least thirteen moons. Three of them, Ganymede, Callisto and Io, are larger than ours. Beyond Jupiter, at between 9·0 and 10·1 AU, is the ringed planet, Saturn. It is slightly smaller than Jupiter. One of its ten known moons, Titan, is larger than the planet Mercury. The diameters of the second pair of giants, Uranus and Neptune, are both about four times that of the earth. While Uranus, whose system of narrow rings was discovered in March 1977, orbits the sun at between 18·3 and 20·1 AU, Neptune's orbit, like that of Venus, is nearly circular. Its distance from the sun is always within one per cent of 30·1 AU. Uranus has five small moons. Triton, one of Neptune's two lunar companions, is the largest in the solar system. Taken together, the combined volume of the four giant planets is just over one thousand times that of the four inner planets but their combined weight is only a little more than two hundred times greater.

At the time of writing, the outermost planet was Pluto. But like that of Mercury, the orbit of this Mars-sized world is very elliptical and between 1979 and 1999 Pluto will lie within 29·6 AU of the sun, which is inside the orbit of Neptune. At its remotest – 49·3 AU – it grazes the extra-planetary fringes of the solar system where millions, perhaps billions, of comets are orbiting at distances of up to 100,000 AU.

Planets and moons

The diagram below shows the nine planets and their thirty-three known moons. The planets are ranged, left to right, in order of their distances from the sun. They have been drawn to scale and their sizes may be compared to that of the sun, the edge of which is shown beneath them.

Their true diameters – measured in kilometres – and other data are given in the table on the right. Oblateness is a measure of how well a planet's shape compares with a perfect sphere (○). The mass, volume and surface gravity of each planet are compared with that of the earth. Surface or cloud-top temperatures are given in degrees Centigrade.

The moons are arranged in order of their distances from their parent planets. The larger moons are drawn to scale but the sizes of the smaller ones have been exaggerated to make them visible. An irregular chunk of rock measuring about 15 by 11 by 12 kilometres, the Martian moon Deimos is one of nine known moons with estimated diameters of less than 50 kilometres. At the other end of the scale, five of the larger moons are bigger than the earth's (diameter, 3,476 kilometres).

Earth
1 Moon

Mars
2 Phobos
3 Deimos

Jupiter
4 Amalthea
5 Io
6 Europa
7 Ganymede
8 Callisto
9 Leda
10 Himalia
11 Lysithea
12 Elara
13 Ananke
14 Carme
15 Pasiphae
16 Sinope

Saturn
17 Janus
18 Mimas
19 Enceladus
20 Tethys
21 Dione
22 Rhea
23 Titan
24 Hyperion
25 Iapetus
26 Phoebe

Uranus
27 Miranda
28 Ariel
29 Umbriel
30 Titania
31 Oberon

Neptune
32 Triton
33 Nereid

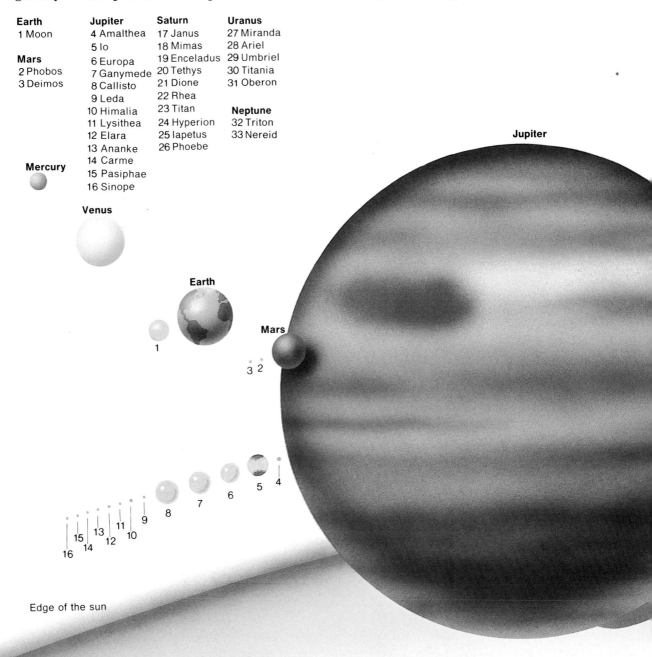

Mercury

Venus

Earth

Mars

Jupiter

Edge of the sun

Symbol	Planet	Equatorial diameter	Oblateness	Mass	Volume	Density	Surface temperature	Surface gravity	Moons
☿	Mercury	4,880	0	0·055	0·06	5·4	Day 350° Night−170°	0·37	0
♀	Venus	12,104	0	0·815	0·88	5·2	480°	0·88	0
⊕	Earth	12,756	0·003	1	1	5·5	22°	1	1
♂	Mars	6,787	0·009	0·108	0·15	3·9	−23°	0·38	2
♃	Jupiter	142,800	0·06	317·9	1,316	1·3	Cloud−150°	2·64	13
♄	Saturn	120,000	0·1	95·2	755	0·7	Cloud−180°	1·15	10
♅	Uranus	51,800	0·06	14·6	67	1·2	Cloud−210°	1·17	5
♆	Neptune	49,500	0·02	17·2	57	1·7	Cloud−220°	1·18	2
♇	Pluto	6,000	Unknown	0·1	0·1	Unknown	−230°	Unknown	Unknown

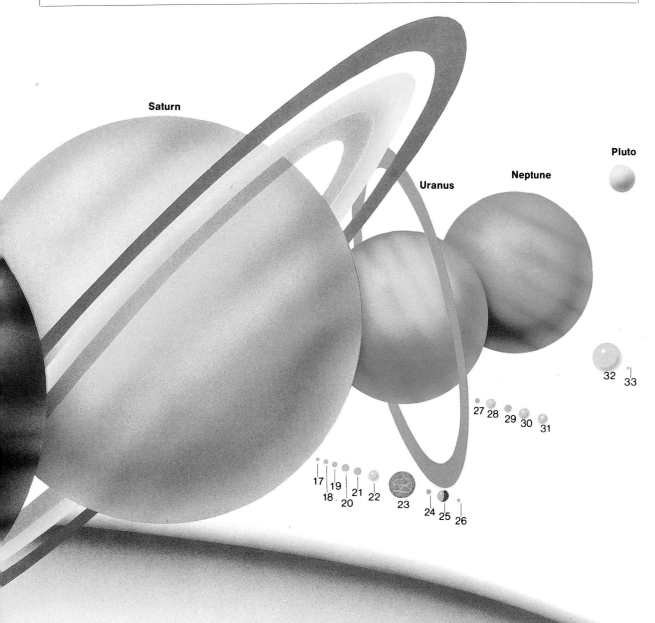

Saturn

Uranus

Neptune

Pluto

17 18 19 20 21 22 23 24 25 26 27 28 29 30 31 32 33

Symbol	Planet	Maximum distance from the sun	Minimum distance from the sun	Average distance from the sun		Eccentricity of orbit	Inclination of orbit	Orbital velocity	Orbital period years	Orbital period days	Inclination of axis	Rotation period days hours minutes
☿	Mercury	69·7	45·9	57·9	*0·387*	0·206	7°	47·9		88	<2°	58 16
♀	Venus	109	107·4	108·2	*0·723*	0·007	3° 24′	35		224·7	3°	243 retrograde
⊕	Earth	152·1	147·1	149·6	*1*	0·017	0°	29·8		365·26	23° 27′	23 56
♂	Mars	249·1	206·7	227·9	*1·524*	0·093	1° 51′	24·1		687	23° 59′	24 37
♃	Jupiter	815·7	740·9	778·3	*5·203*	0·048	1° 18′	13·1	11·86		3° 05′	9 50
♄	Saturn	1,507	1,347	1,427	*9·539*	0·056	2° 29′	9·6	29·46		26° 44′	10 14
♅	Uranus	3,004	2,735	2,870	*19·18*	0·047	0° 46′	6·8	84·01		82° 05′	20±3 retrograde
♆	Neptune	4,537	4,456	4,497	*30·06*	0·009	1° 46′	5·4	164·8		28° 48′	21±2
♇	Pluto	7,375	4,425	5,900	*39·44*	0·25	17° 12′	4·7	247·7		unknown	6 9

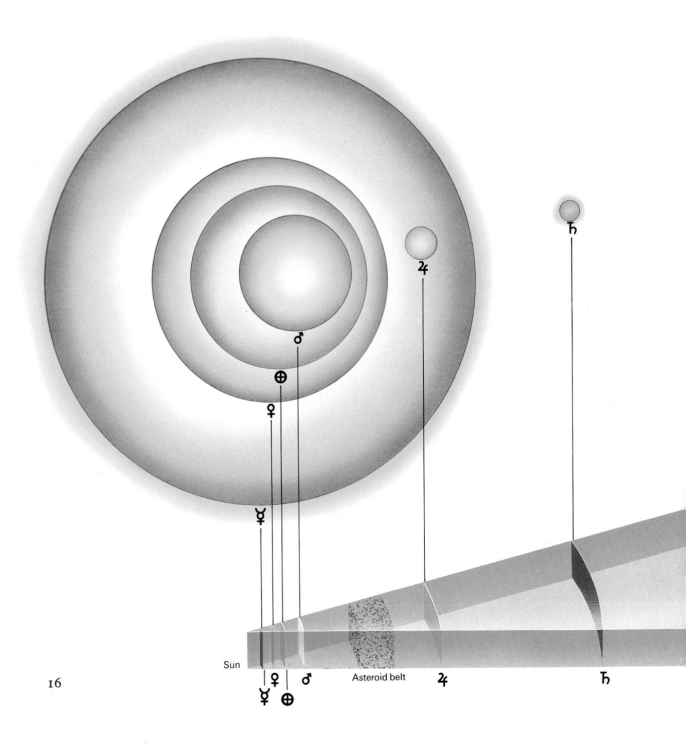

Sun ☿ ♀ ⊕ ♂ Asteroid belt ♃ ♄

Orbits and rotations

The lower part of this diagram shows a slice through the orbits of the planets, which are arranged, left to right, according to their distances from the sun. The true average, maximum and minimum distances, measured in millions of kilometres, are given in the table at the left. Set in italics, the average distances are also given in astronomical units, or AU, by which scale the earth–sun distance is unity.

The yellow discs represent the sun. On the left the sun is shown large as it might appear in the 'sky' of Mercury, the closest planet. At greater distances the solar disc becomes progressively smaller. In 1979, Pluto will cross inside the orbit of Neptune. (There is no chance of a collision – their orbits have a flyover–underpass relationship.)

Eccentricity is a measure of how well a planet's orbit compares to a perfect circle (○). The values for orbital velocity are given in kilometres per second. Inclination is a measure of the tilt of a planet's orbital plane with respect to that of the earth. Rotation periods are measured with respect to the stars.

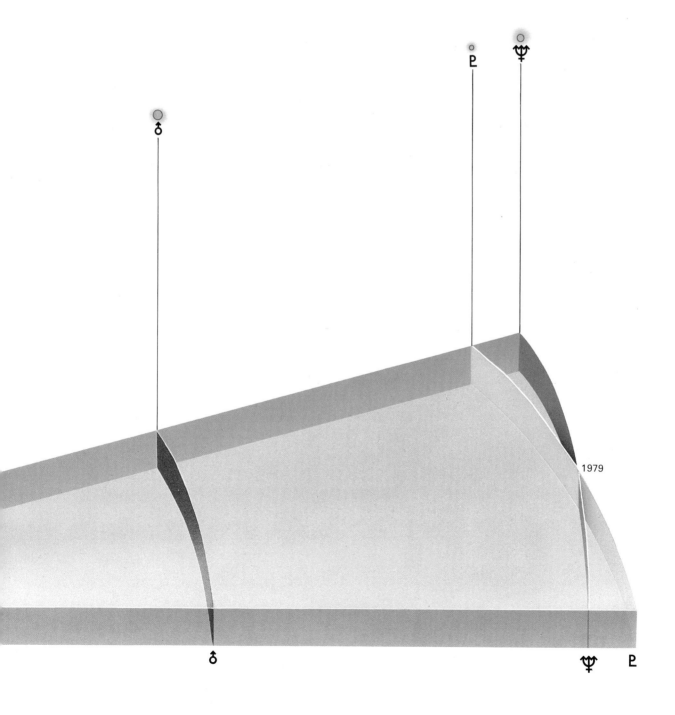

1979

Interiors and atmospheres

The diagram below, which is not to scale, shows what the solar system is made of. Apart from empty space, its commonest ingredient is hydrogen. Seventy-one per cent of the sun (which accounts for 99·9 per cent of the total mass of the solar system) is hydrogen. The other major solar ingredient – 27 per cent – is helium. In the sun, hydrogen and helium form a plasma – a fourth state of matter which is neither solid, liquid nor gas but a very hot and dense mass of nuclei and electrons.

In a cooler but still dense state, 'metallic' hydrogen surrounds the rocky cores of Jupiter and Saturn. It is called metallic because it behaves like a liquid metal, producing a magnetic field as does the liquid metallic core of the earth.

The cores of the giant planets, the interiors of the inner planets – Mercury, Venus, Earth, Mars – and many asteroids and meteoroids are composed of a rocky mix of oxygen, silicon, aluminium, iron and other elements. The cores of the inner planets, large asteroids like Ceres and some meteoroids are metallic.

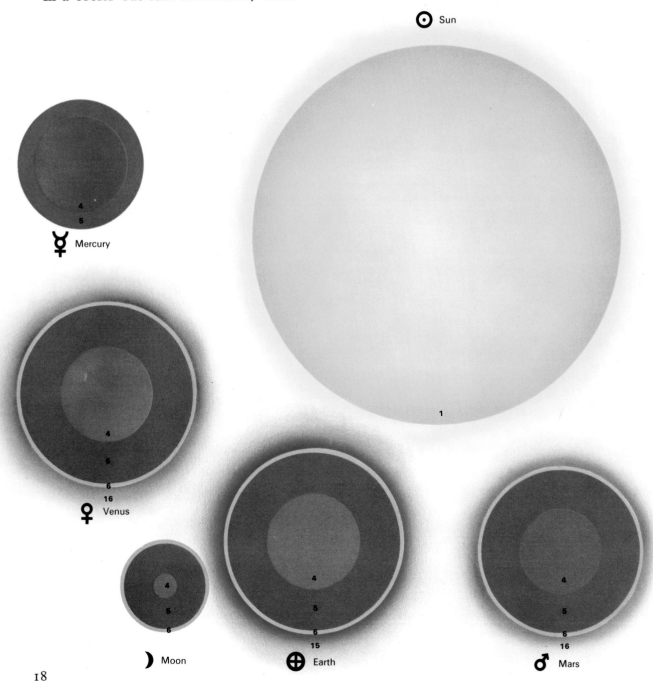

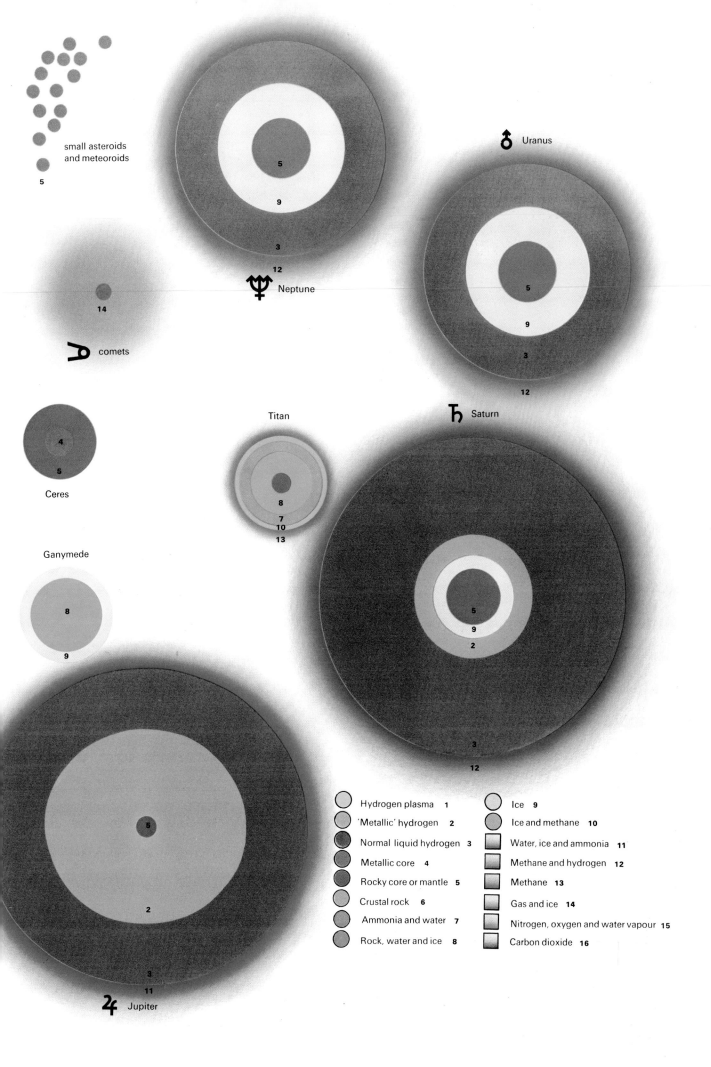

small asteroids
and meteoroids

5

14

comets

Neptune

Uranus

Ceres

4
5

Titan

Saturn

8
7
10
13

Ganymede

8
9

5
9
2

3
12

5
9
2

2

3
11

Jupiter

Hydrogen plasma 1
'Metallic' hydrogen 2
Normal liquid hydrogen 3
Metallic core 4
Rocky core or mantle 5
Crustal rock 6
Ammonia and water 7
Rock, water and ice 8

Ice 9
Ice and methane 10
Water, ice and ammonia 11
Methane and hydrogen 12
Methane 13
Gas and ice 14
Nitrogen, oxygen and water vapour 15
Carbon dioxide 16

SUN

The sun is a luminous sphere of very hot gas. Together with its retinue of planets and moons, it travels around the centre of the Milky Way in a spiral stream of similar stars. This stream forms one of the arms of the spinning galaxy. At the sun's distance from the centre, 30,000 light years, one complete revolution takes 200 million years, at a speed of 290 kilometres a second. The sun is also moving among its neighbours, drifting away from the stars around Canopus towards those around Vega at a rate of some nineteen kilometres a second. For viewers at mid-northern latitudes, its destination lies near zenith at midnight in June.

The equatorial diameter of this itinerant globe of energy is 1,392,000 kilometres. The polar diameter is only some seventy kilometres less than the equatorial, so the sun is a near perfect sphere. Its visible surface, whence heat and light are radiated into space, is a seething patchwork made up of millions of cloud tops. The sun is yellow because the average temperature of these clouds is 6,000 °K. If they were hotter, the sun would appear more blue. If they were cooler, it would be redder. The clouds contain high-temperature hydrogen and helium and belong to a layer, between one hundred and four

---

The painting on the left shows the sun as it might appear when viewed from near-earth space. Above our light-scattering atmosphere, both the bright photosphere and the relatively pale fire of the corona – the tenuous outer shell of the sun's atmosphere – can be seen. The moon and the Milky Way are in the background.

hundred kilometres thick, called the photosphere. The life-span of an individual photosphere cloud is between seven and ten minutes; in this time its diameter may grow to more than 1,400 kilometres. A 1,800-kilometres-an-hour updraught lifts the hot light-releasing gas from the base to the top of the cloud, where it cools and falls back down the sides. As a mature cloud contracts and fades, its place is taken by fresh material from below.

This fluctuating network of clouds gives the solar surface a mottled appearance which astronomers call granulation. Each granule corresponds to a bright polygonal cloud separated from its neighbours by intergranular lanes of cooler, and therefore darker, gas. A temperature drop, 10,000 °K to 4,200 °K, from cloud base to top is also one of the reasons why the edge of the sun, or limb, appears darker than the centre. When we view the middle of the sun's disc, we are looking directly down through the solar atmosphere at light which is being radiated by hot gas from deeper, brighter regions of the photosphere. When we view the edge of the sun, we are looking through a much greater thickness of solar atmosphere, which has a filtering effect, and we only see the light coming from the upper, cooler, and therefore darker part of the photosphere.

From time to time, the pulsating pattern of cloud cells is invaded by vast depressions whose floors are cooler, and therefore darker, than the surrounding photosphere. These are called sunspots. One of the largest, recorded in 1858, measured 225,000 kilometres across, nearly twenty times the radius of the earth. The central

Solar structure

This diagram shows the sun (right), a slice through it (below right) and an enlarged detail of a small part of the surface (below left).

As well as heat and light, the sun (right) is the source of a non-stop broadcast of atomic particles, represented here by concentric rings. Called the solar wind, this invisible tenuous stream of matter expands through the visible and equatorially extensive outer part of the sun's atmosphere, the corona (far right), to reach all parts of the solar system.

The solar wind 'blows' across the interplanetary vacuum at about 500 kilometres a second – a gentle breeze compared to the 2,000-kilometre-a-second gales of protons and electrons emitted by active zones during flare storms.

The detail (below left) shows a small segment of the sun's gaseous surface and the convective zone beneath it. Above a layer of cushion-like light-producing photosphere clouds, in which sunspots may occur, a forest of chromosphere spicules reaches up into the corona.

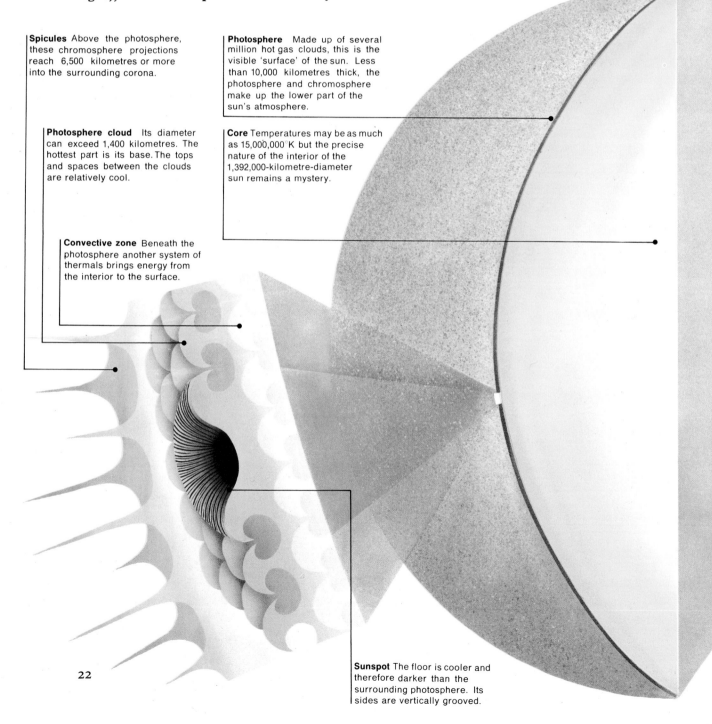

Spicules Above the photosphere, these chromosphere projections reach 6,500 kilometres or more into the surrounding corona.

Photosphere cloud Its diameter can exceed 1,400 kilometres. The hottest part is its base. The tops and spaces between the clouds are relatively cool.

Convective zone Beneath the photosphere another system of thermals brings energy from the interior to the surface.

Photosphere Made up of several million hot gas clouds, this is the visible 'surface' of the sun. Less than 10,000 kilometres thick, the photosphere and chromosphere make up the lower part of the sun's atmosphere.

Core Temperatures may be as much as 15,000,000 K but the precise nature of the interior of the 1,392,000-kilometre-diameter sun remains a mystery.

Sunspot The floor is cooler and therefore darker than the surrounding photosphere. Its sides are vertically grooved.

Solar wind The sun's outer atmosphere expanding into space.

Sunspots and flares During active phases of the solar cycle the surface of the sun becomes patchy. Very bright transient flares contrast with dark sunspots.

Corona The hottest part of the sun's atmosphere but very rare and not therefore as bright as the photosphere.

Charged particles Emitted in streams during solar flare surges. They are set spiralling across interplanetary space by the rotation of the sun.

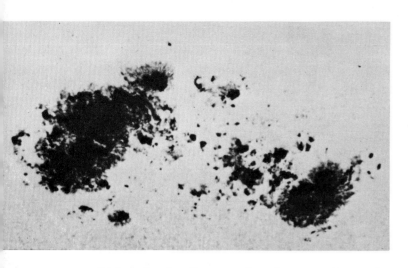

Taken in April 1947, the photograph on the right shows the sun during the active phase of an eleven-year solar cycle (see diagram on page 26). The prominent group of sunspots in the southern hemisphere, which is also shown in the close-up above, covers an area of over fourteen billion square kilometres – nearly thirty times the surface area of the earth.

The photograph below, taken through a telescope hoisted by balloon to an altitude of twenty-seven kilometres, shows another sunspot group. The magnetically grooved rim, or penumbra, of the large spot in the left-hand side of the picture shows up well, as does the granular appearance of the photosphere. Each granule is a short-lived polygonal cloud of hot bright gas separated from its neighbours by lanes of cooler darker gas.

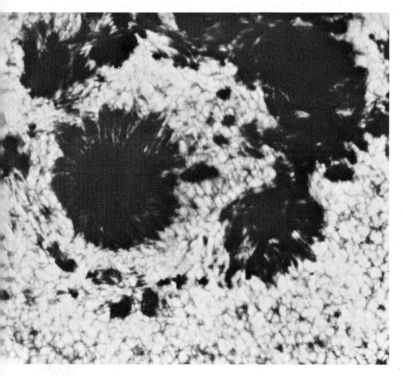

floor of a sunspot, or umbra, which appears dark in photographs, is less than three quarters as hot and less than one quarter as bright as the adjacent photosphere. The umbra is bounded by the penumbra, the vertically grooved sides of which slope up and out to meet the surrounding surface. The alignment of these peripheral grooves forms a pattern familiar to anyone who has sprinkled iron filings on a sheet of paper placed above one pole of a bar magnet. Magnetically, sunspots are either north- or south-seeking. Where they appear in pairs, one member of each pair is north-seeking and the other is south-seeking.

The magnetic nature of sunspots has led to a suggestion as to how they are formed. Beneath the yellow visible surface of the sun, it is proposed, lies an invisible network of magnetic channels arranged like a loosely wound ball of coiled and somewhat springy wire. Where a loop of this network breaks the surface, a pair of sunspots is produced. One north-seeking, the other south-seeking, they are joined by an invisible magnetic loop arching above the photosphere.

Sunspots were recorded by Chinese astronomers more than thirteen centuries ago but in the contemporary western world, where the existence of such blemishes did not fit too well with the authoritarian rubrics of an earth-centred universe, they were ignored. The solar surface was perfect and smooth, said the Church Fathers, and that was that. Western records of sunspots do not begin until about 1610 when Galileo damaged the sight of one of his eyes looking at them through one of his telescopes. Some two centuries later a German amateur astronomer, Samuel Heinrich Schwabe, was looking for a new planet inside the orbit of Mercury, the elusive Vulcan of science fiction. To achieve this he kept up a near constant watch over the solar surface, in the hope that one day he would spot his quarry in transit. To pass the time he took note of the daily variation in the number of sunspots. In 1843, after seventeen years of continuous observation, he realized that he had made a significant discovery, not the hoped-for planet but an unexpected regular rhythm in his sunspot tallies. Every eleven years, he noticed, the number of sunspots reached a maximum. The number would then decline over

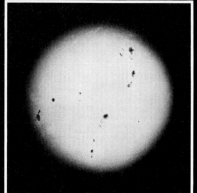

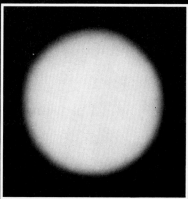

Sunspots were common in 1958 (far left), when the sun was active, and rare in 1952 (near left), when it was not. The large photograph above shows the giant sunspot group of April 1947.

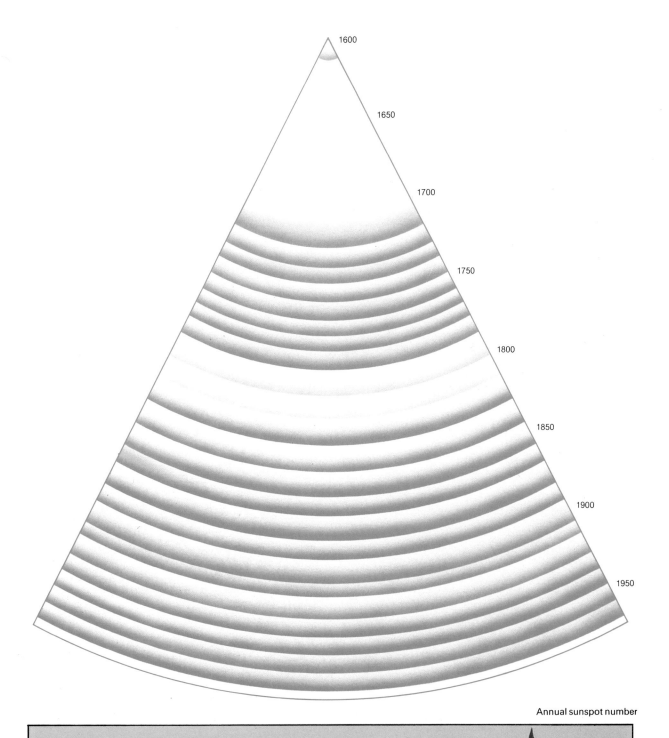

1600
1650
1700
1750
1800
1850
1900
1950

Annual sunspot number

150

100

50

0

1875
1900
1925
1950
1975

the next seven years to a minimum before a four-year build-up to another maximum. This eleven-year solar cycle, which has continued to the present time, was then patiently traced back through the records of sun-watching astronomers to the year 1715. At this point there was a break. Checking back further, only three maxima were found for the whole of the preceding century, together with a marked reduction in the sunspot total for the seventy-year period between 1645 and 1715. Was the sun's behaviour really different in the seventeenth century, or was the record unreliable? An answer came from an unexpected source.

At the beginning of this century an American scientist, Andrew Ellicott Douglass, was investigating the year-by-year variation in the growth of trees. He too found an approximately decennial cycle, reflected in the annual variation of the widths of their growth rings. But when he came to examine cross-sections of timber from the late seventeenth century the rings, he saw, were more or less uniform. The pattern was missing. One wonders what Archimedian exclamation he made on reading, in 1922, a paper by the British astronomer Edward Walter Maunder, commenting on the seventy-year absence of the sunspot cycle at that same period. Subsequent studies have demonstrated other possible effects of the solar cycle. In the winter of 1683 the Thames froze over and frost fairs were staged on the ice. Another cold spell at the beginning of the nineteenth century contributed to Napoleon's reverses on the Russian front. Both these events coincided with periods of sunspot scarcity. But believing the record to be less than completely reliable, astronomers took little more notice of these gaps in the eleven-year cycle until

very recently. In 1977 an American astronomer, John Eddy, published a careful review of the historical evidence which, coupled with data from recent tree-ring studies, indicates that other, longer-term cycles are at work.

European reports of sightings of aurorae, or nocturnal 'lights' – produced by the interaction of particles emitted by an active sun with others in the earth's upper atmosphere – have been compared with records of naked-eye visible sunspots from Chinese, Japanese and Korean archives. The reported appearance of the corona – spiky when the sun is active, reduced when it is not – has been noted from historical accounts of eclipses. In all three cases there are gaps in the record, beginning and ending at the same dates, the same long intervals when aurora, large sunspot and spiky corona sightings became rare, or ceased altogether. These results are also in harmony with those of a set of tree-ring investigations, the nature of which raises the exciting possibility that we may soon be able to trace the sun's history a lot further back into the past.

Carbon 14 – a radioactive form of ordinary carbon 12 – is produced in the earth's atmosphere by the interaction of earthly atmospheric atoms with cosmic ray particles emitted by stars and supernovae from other parts of the galaxy. It is removed from the atmosphere by plants which absorb carbon dioxide. The amount of carbon 14 present in the atmosphere varies with the intensity of cosmic ray bombardment, which is in turn affected by the strength of the sun's magnetic field. An active sun has a strong magnetic field which tends to shield the inner planets from cosmic rays and so reduce the level of carbon 14 production in the earth's upper atmosphere; a quiet sun has a weak magnetic field which lets cosmic rays through. The annual growth rings of trees, therefore, contain variable amounts of carbon 14.

The interpretation of this biological recording of the sun's behaviour suggests that the last 150 years of solar activity represents an approaching peak in a much longer-term pattern than that of the eleven-year cycle. Since 3000 BC, according to the bristlecone pine trees, there have been six eras of activity (including the present one) interspersed with six much longer eras of calm (including the 1645–1715 interval). When the Ancient Greeks wrote their accounts of the

The lower of the two diagrams opposite shows a hundred years of peaks and troughs in the annual number of recorded sunspots. The average interval between troughs is 11.2 years. The upper diagram shows how the cycle can be traced back to the beginning of the seventeenth century when the invention of the telescope enabled observations to begin. There are two gaps in this record. During the whole of the longest – the seventy-year interval from 1645 to 1715 – the total number of sunspots seen was less than what is recorded in an average single year today.

Total eclipse, 1970. In the upper part of the photograph, a minute chink of photosphere creates the 'diamond ring' effect. The rest of the surrounding light is that of the solar corona.

A total eclipse occurs when the moon passes directly between the sun and the earth. The duration of the eclipse and the width of the moon's shadow passing over the earth's surface depends on the sun–earth and earth–moon distances at the time. An eclipse occurring while the earth is far from the sun (in the middle of the northern hemisphere summer) and the moon is close to the earth (once a month) will cast a wider shadow and last longer than at other times. Thus the longest total eclipse of modern times, that of June 1973, lasted 7 minutes and 14 seconds. Under different conditions, that of October 1986 may only last for one second.

To gain observing time, astronomers may load their instruments aboard supersonic aircraft which can keep pace with the moon's shadow as it races across the face of the earth.

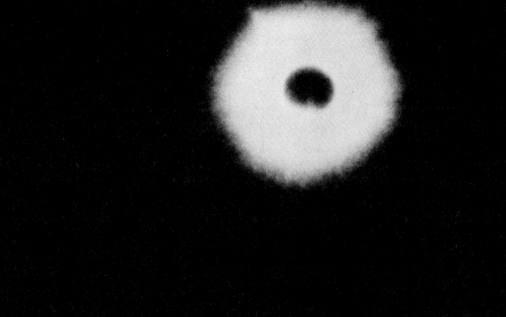

The March 1970 eclipse at totality. The photosphere is now completely obscured and, in the absence of that bright light source, the pale glow of the solar corona is more apparent than it is in the photograph on the left.

The size and appearance of the corona varies with the level of solar activity. This eclipse occurred when the sun was close to a peak in the eleven-year sunspot cycle. The sun's outer atmosphere is extensive and spiky. When sunspot activity is reduced it is less so. Thus historical descriptions of the appearance of the corona during eclipses are important clues to the past behaviour of the sun. The Egyptians, for example, often represented the sun with wings, which fits neatly with modern tree-ring evidence that the sun was active when the pyramids were built. But in those accounts of eclipses between 1645 and 1715 where the corona is mentioned at all it is reported as being no more than a thin band of dull light – a description consistent with the concurrent recorded absence of sunspots.

Taken at six-minute intervals, this sequence of photographs shows the progress of a partial eclipse over Washington in July 1972. At the beginning of the eclipse (top right) the moon's disc cuts into that of the sun.

Insets show the solar corona as it appears during a total eclipse when the sun is relatively active. Its spiky structure is due to the shaping of the sun's upper atmosphere by its magnetic field. The patterns are reminiscent of those produced by sprinkling iron filings on a piece of paper suspended above a bar magnet.

The corona is not normally visible from earth because light originating from the much brighter photosphere is scattered into the field of view by the earth's atmosphere. To overcome this problem, astronomers use the coronagraph, an instrument in which a metal disc produces an artificial eclipse.

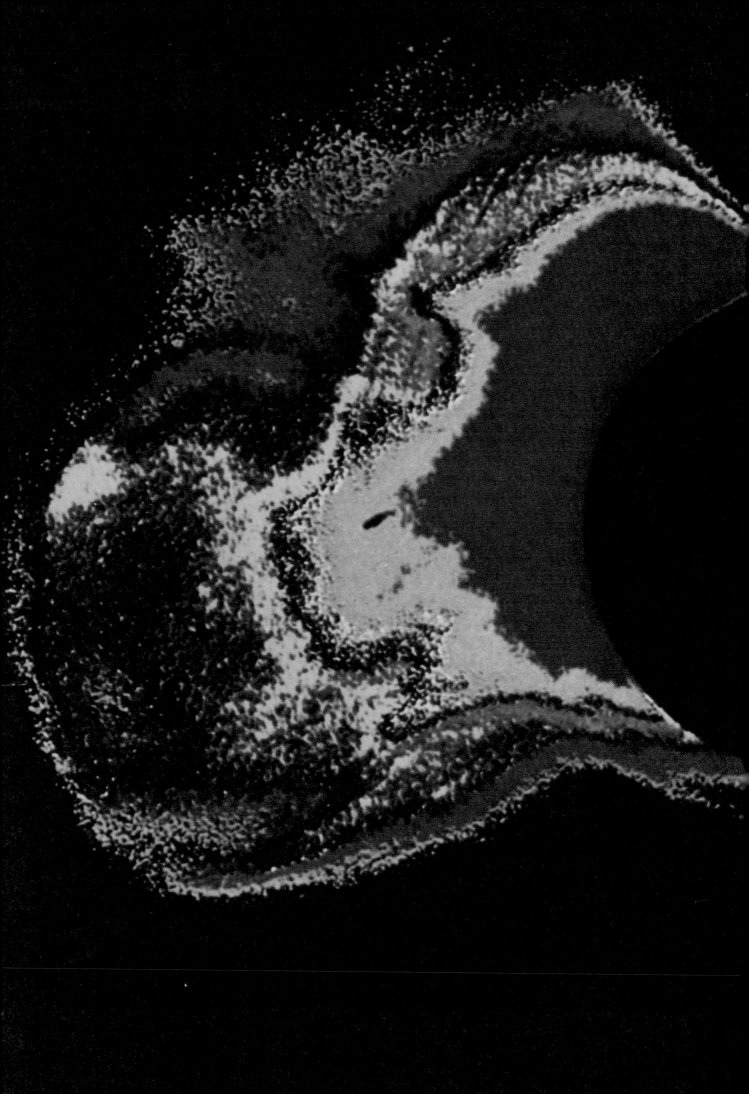

Colour-coded by a computer, this originally black and white image of the sun's
corona, taken through a coronagraph
aboard an earth-orbiting Skylab spacecraft, shows variations in its density.
Extensive near the equator but reduced
around the poles, the sun's outer atmosphere thins out with altitude where it
boils off into space.

Inset right, a photograph taken from
an aircraft flying at an altitude of 12 kilometres shows the corona during the total
eclipse of May 1965.

heavens the sun was quiet. There were almost certainly no sunspots to be seen and none was recorded in the writings of Aristotle, which is one reason why the Vatican, whose observational astronomy was more or less confined to scanning classical manuscripts, opposed the truth of Galileo's discovery.

Once a month the moon passes between the sun and the earth but, because its orbit lies in a plane which is slightly tilted with respect to that of the earth's orbit around the sun, the moon's shadow does not always fall upon our planet. When it does, we have an eclipse.

When only part of the lunar silhouette crosses the face of the sun, the eclipse is described as partial. During an annular eclipse the sun, moon and earth are in perfect alignment but, because the moon is at a relatively distant point from the earth along its slightly elliptical orbit, its disc appears smaller than that of the sun and a ring of photospheric light shines around its edges. By a convenient natural coincidence, the sun's diameter is 400 times larger than the moon's and, when the latter is relatively close to the earth, exactly 400 times further away. Under these conditions, an eclipse is total. The lunar silhouette slides neatly over the photosphere and for a few moments it is sometimes possible to catch a glimpse of the sun's comparatively faint upper atmosphere.

Total eclipses are rare and short-lived events. There are only seventeen opportunities to see one between 1977 and the end of the century from various vantage points around the globe – good news for travel agents but less so for astronomers, who even use supersonic aircraft to follow the path of the moon's shadow as it races across the earth's surface. A less spectacular but more practical way of observing the sun's upper atmosphere is to use a coronagraph, an instrument invented in 1930 by the French astronomer Bernard Lyot which simulates an eclipse by

Active sun. Rooted in active zones of the photosphere, enormous fiery spindrifts of solar gas leap from behind the lunar silhouette during a total eclipse. These pale flames climb invisible magnetic lines before falling back to the solar surface. To give some idea of their size, a photograph of the earth (diameter, 12,756 kilometres) to the same scale has been superimposed.

34

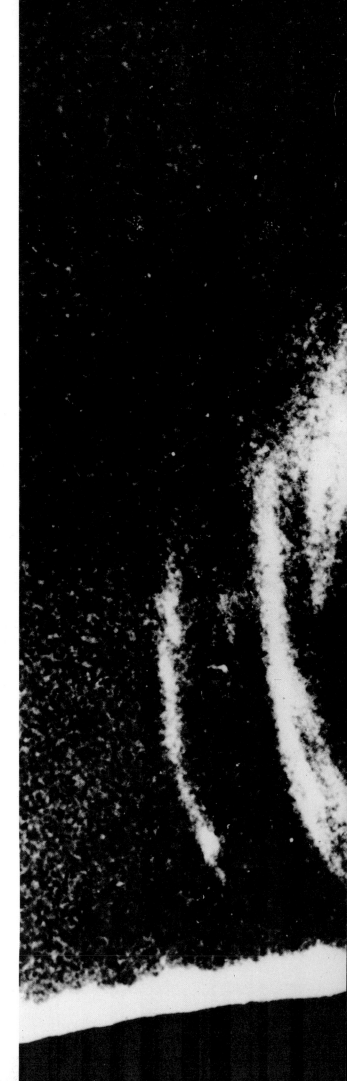

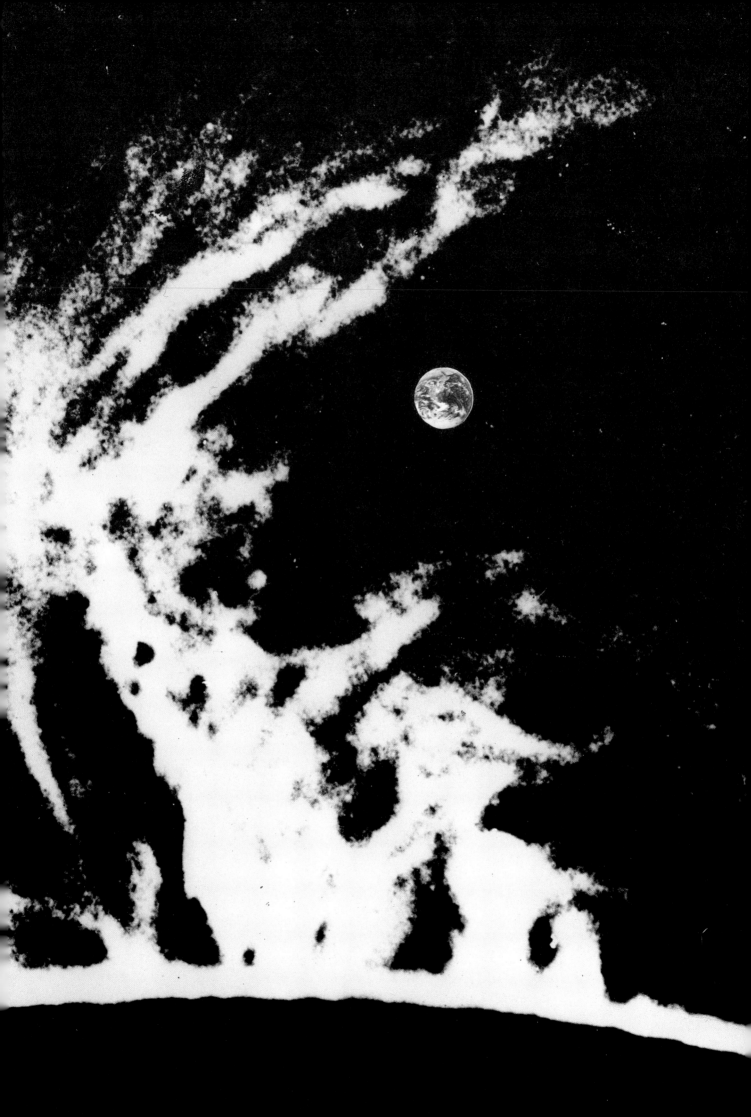

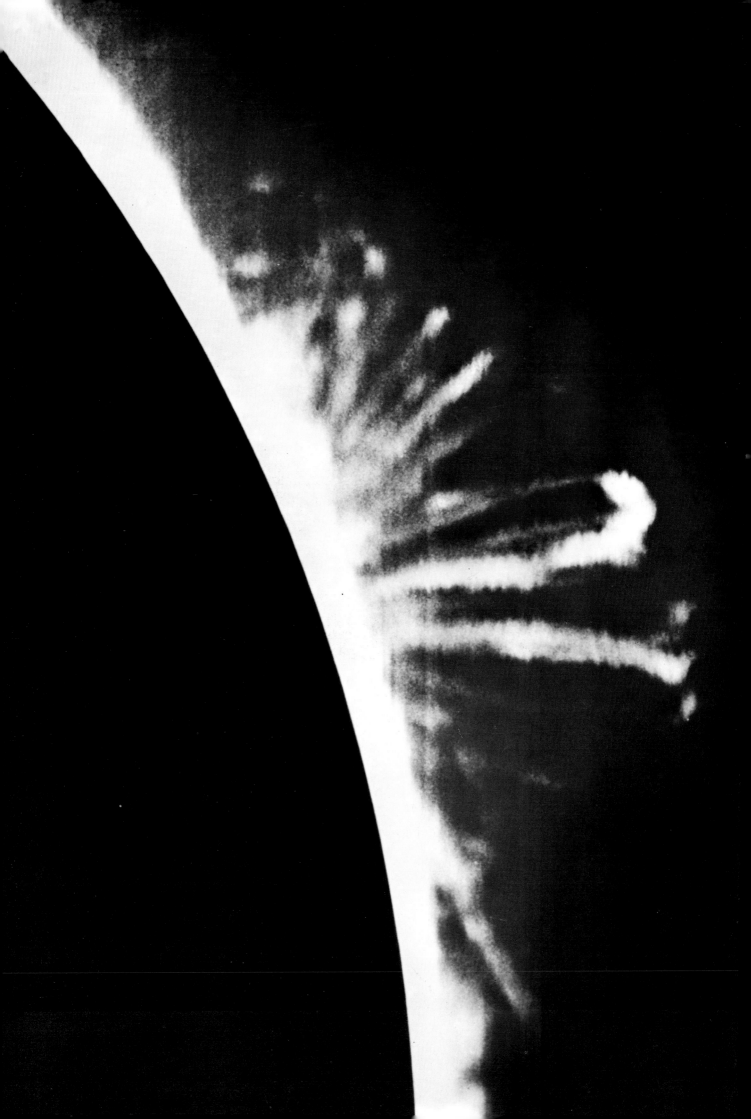

Surrounded by swirling chromosphere spicules, a typical active zone (above) contains sunspots, bright plages and long dark filaments. Above another active centre (left), a spiky eruption reaches up into the corona. The mottled appearance of the sun's surface (right) reflects the irregular distribution of active zones which show up as bright patches. The twisting sheet of gas on the left of the sun loops out more than a third of a million kilometres into space. This extreme ultraviolet photograph – recorded by the light of ionized helium gas, one of the sun's principal components – was taken from earth orbit aboard the manned Skylab 4 spacecraft.

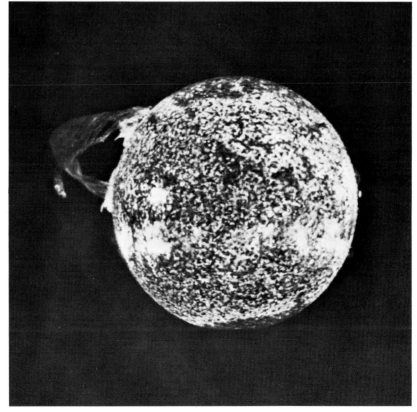

A computer-generated colour photograph of a burst of solar energy. The black and white image (above), taken by Skylab 2 astronauts, depicts a similar event – an eruption of a 370,000-kilometre streamer of helium gas from the solar surface. In the colour photograph, different hues reflect different intensities of ultraviolet emission as recorded by a black and white image. Low intensities, dark greys in a black and white photograph, are represented by red, yellow and green. High values, light greys in black and white, become blues, violet and white.

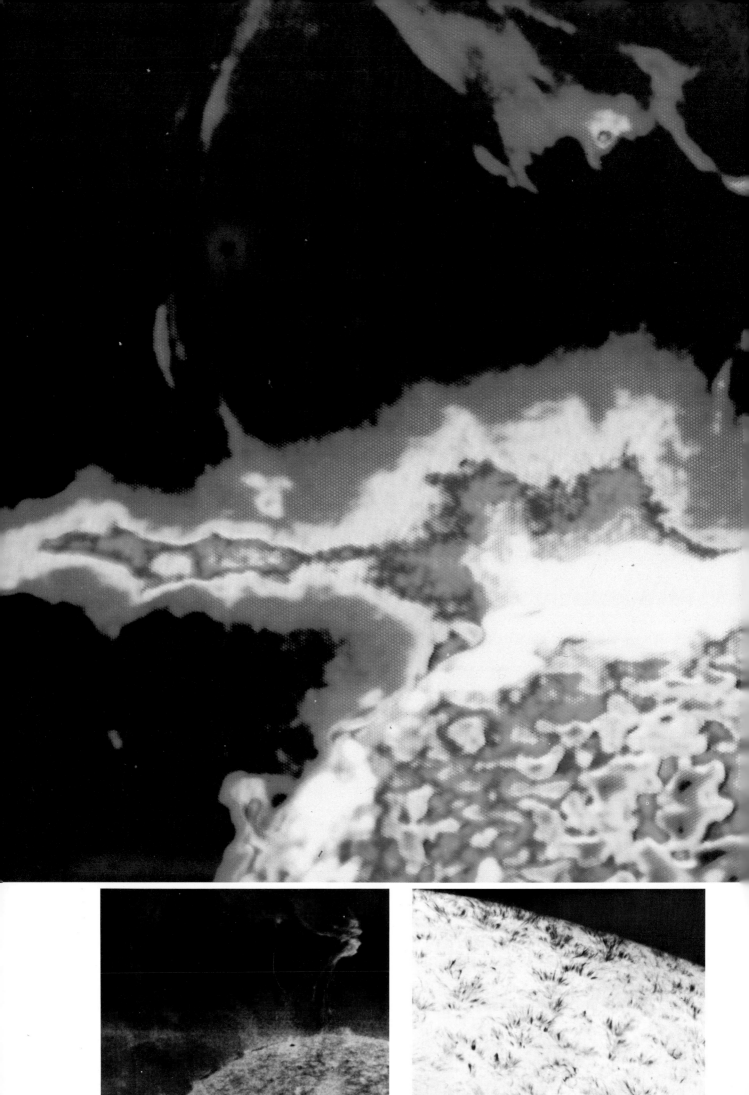

blocking out light coming from the photosphere.

In addition to the cooler darker sunspots, the photosphere may also be masked or marked by transient regions of different temperature and therefore brightness. The largest of these are called plages. Less extensive areas which look like tufts of wool are called flocculi. Depending on their temperatures, plages and flocculi may be brighter or darker than the adjacent photosphere. Large irregular patches, on average 10 per cent brighter than their surroundings and normally located near sunspot groups, are called faculae, or torches. Faculae often form the photospheric roots of bright prominences which may extend a million kilometres or more above the solar surface. Sudden intense surges in the brightness of very localized patches above the photosphere are called flares.

There are two atmospheric layers above the photosphere. The lower is the chromosphere, up to 8,000 kilometres thick, named for its reddish hue. The upper, called the corona, is the pale fire envelope which extends several solar diameters beyond the sun's bright surface.

Compared with the huge volume of the main body of the sun, the chromosphere is a mere onion skin. From its lower regions, some 1,500 kilometres thick, grows a transparent forest of gaseous projections called spicules. Like the cloud cells of the underlying photosphere, these are also transient in nature. An individual spicule lasts only fifteen minutes, but at any given moment 100,000 of them form a spherical field of corn-like stems of gas reaching up into the corona above. Developed from bases 500 kilometres wide, they extend upwards at between 20 and 30 kilometres a second to heights of up to 8,000 kilometres above the solar surface. Temperatures within a spicule column vary from 10,000 °K in the middle to 50,000 °K at its edges. Although it is much hotter, the chromosphere is less dense and therefore less bright than the photosphere. In the corona temperatures reach 2,000,000 °K, but its pearly light is only as bright as the light of the full moon. Coronal gases are constantly boiling off, filling interplanetary space with an outward flow of electrically charged particles called the solar wind.

It is normally very difficult to observe the sun's pale outer atmosphere from the surface of the earth because our atmosphere scatters the

Another black and white image of a solar eruption (far left), colour coded (above) to reveal ultraviolet intensity and therefore density variations, as it enters the hot, but tenuous (and therefore here black), corona. The third photograph (near left) shows spicules in the chromosphere.

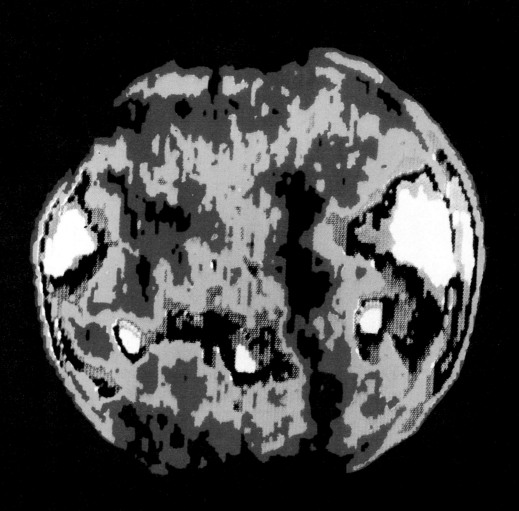

field of view with light from the much brighter photosphere. Thanks to the space age, astronomers can now launch their instruments beyond the earth's atmosphere where they can also survey the entire range of solar radiation, only 40 per cent of which is in the form of the visible rainbow bands of sunlight. The rest is at other wavelengths, many of which are unable to penetrate the earth's atmospheric filter. These include the relatively shorter waves of sun-tanning ultraviolet, X-rays and gamma rays, and the comparatively longer waves of warmth-giving infra-red and the radio spectrum. One of the first results came from the surface of the moon when an unmanned pre-Apollo Surveyor spacecraft confirmed the considerable extent of the solar corona. Sensors directed at the lunar horizon recorded the coronal glow rising some time before and setting some time after the more visible disc of the sun.

Between May 1973 and February 1974 three manned post-Apollo Skylab missions logged thousands of hours of solar observation from earth orbit, many of them in hitherto hidden wavebands. Photographs taken from space in ultraviolet light have enabled astronomers to construct temperature maps of the sun and its atmosphere. They have revealed, in much greater detail than was possible before, the forms of prominences, arches, filaments and fountains of solar material which appear in the chromosphere and corona during periods of magnetic activity. X-ray images show the uneven distribution of the coronal envelope, reflecting the uneven nature of solar surface activity, which appears to be generally depressed at the poles. However, while gaps in the corona, called coronal holes, are sites of reduced X-ray emission and lower temperatures on ultraviolet maps, they are thought to be strong sources of solar wind.

The years between maxima in the sunspot cycle are sometimes called years of the quiet sun. When sunspots are on the increase, the sun is described as active. Typically, an active centre is heralded some days before the appearance of sunspots by a localized disturbance in the magnetic field just below the surface of the photosphere. At the start of a new cycle of solar activity, this will happen around latitude 30° in either hemisphere. The disturbance is followed by the development of faculae, irregular patches

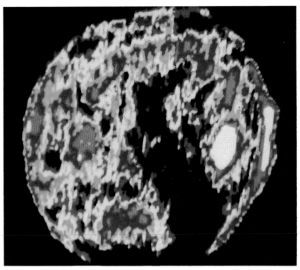

These colour maps of the solar atmosphere were created by computer processing of data from an unmanned American Orbiting Solar Observatory satellite and a manned Skylab mission. On the left, regions of greatest activity, and therefore greatest intensity, as recorded in the X-ray–ultraviolet range of the sun's spectrum, are shown in white. Violet, blue, green, yellow and red patches represent progressively more inactive areas. The least active regions (black) are areas of reduced temperature, pressure and magnetic field strength. These conditions suggest an outflow of material from the solar atmosphere. Coronal holes, as these dark areas are called, may be primary sources of solar wind.

The two photographs on this page were recorded aboard Skylab 3 in August 1973. The first (above) shows a coronal hole extending across the solar surface from the south pole. The second (below) shows changes in its shape and position a day later. The new position is due to the sun's rotation.

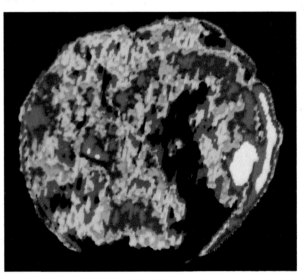

of brilliance in the photosphere from which a ray or arch-shaped extensions will begin to grow through the chromosphere and into the corona. Two or three days later the first sunspot appears, together with further smaller atmospheric projections in its immediate vicinity. Magnetically, this spot will either be north- or south-seeking. (If it is north-seeking, then the first spot to appear as a result of a similar disturbance in the opposite hemisphere will be south-seeking. At the start of the next solar cycle, the situation is reversed. So, if this magnetic reversal is taken into account, the true length of a solar cycle is twenty-two years.) A second spot, of opposite magnetic sign, soon follows the first and numerous smaller spots form between them. After ten days the sunspots reach their maximum size. At this stage there are usually several flare surges, during which clouds of charged particles are exploded from the surface and out into interplanetary space.

Life-spans of active areas vary greatly. Some last a few days, others persist for six months or more. After the last spot – normally the first to appear – has vanished, flare activity is reduced but the bright facular areas continue to expand.

As a solar cycle progresses, there is a tendency for the location of successive active regions to drift towards the equator. By the end of the cycle they are normally forming around latitude 5°. It is not uncommon for cycles to overlap. The last, equatorial sunspots of an ageing cycle may still be visible when the first crop of a fresh cycle begins to appear at higher latitudes.

The gas which makes up solar prominences is ten to one hundred times cooler than the surrounding coronal gas. It owes its visibility to its density. Plumes, loops, arches and other shapes reflect magnetic patterns which control them.

An X-ray photograph, taken during a Skylab mission, of the solar corona, the hottest part of the sun's atmosphere. Just as yellow is the characteristic visible colour 'signature' of the 6,000°K photosphere, so radiation from the X-ray region of the solar spectrum is the characteristic signature of the corona, where temperatures reach several million degrees. Active areas (yellow) and coronal holes (dark) can be seen.

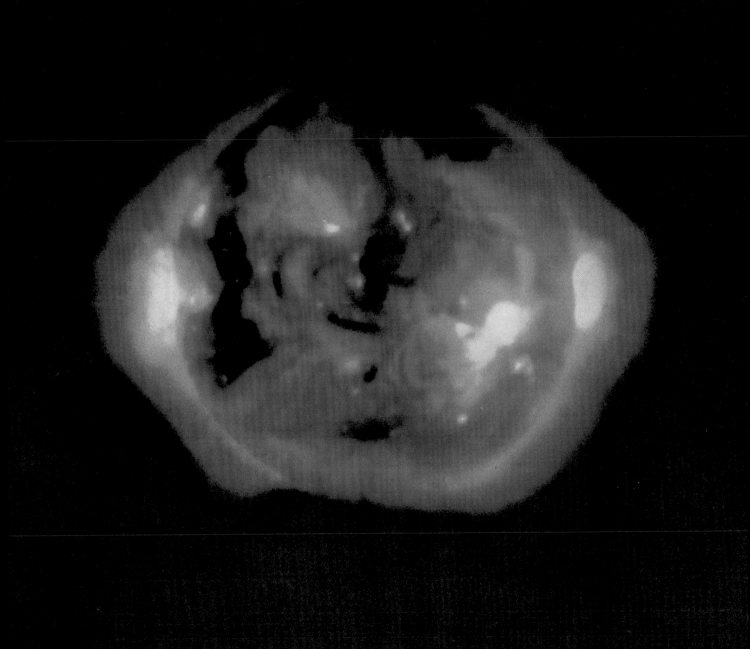

Where these vast pink clouds are a result of condensation of coronal material, earth-sized drops of glowing gas may be seen raining down through the atmosphere. Others are of eruptive origin, the result of millions of tons of material being thrown a distance of a solar diameter or more at speeds of 500 kilometres a second. These eruptions are often linked with shock waves from flare explosions which speed through the solar atmosphere at 1,000 kilometres a second, investing otherwise quiescent bodies of gas with explosive energy.

Because the sun is made out of gas, its rotation period is free to vary with latitude. At the equator one complete turn takes twenty-five days. Nearer the poles the period is about nine days longer. Because their sources are rotating with the photosphere, the intense blasts of charged particles emitted during solar flares are spread about the solar system like water from the spinning nozzle of a lawn sprinkler. Gusting earthwards, these high-energy electrons and protons interact with atoms and molecules in the earth's upper atmosphere, kindling the nocturnal glow of the polar aurorae and disrupting radio communications. When space flights coincide with active phases of the solar cycle, flare particles may endanger the safety of astronauts, but because they travel at a mere 2,000 kilometres a second, only a fraction of the speed of light (300,000 kilometres a second), they take several hours to reach earth distances and their time of arrival can be predicted by careful monitoring of the sun's surface.

Beneath the photosphere, a system of thermals 500 kilometres deep, called the convective zone, brings energy from the interior to the surface. This energy comes from reactions between the atomic nuclei of the elements which make up the sun. The commonest reactions involve the nuclei of hydrogen and helium. The first step in the chain reaction which produces most of the sun's energy is the combination of four ordinary hydrogen nuclei, each consisting of a single proton, to produce two heavy hydrogen, or deuterium, nuclei. Each of these is made up of one proton and one neutron. Changing protons into neutrons releases energy. More energy is released when another proton is added to each of the nuclei, which then become those of a lightweight variety of helium called helium 3. Energy is

Hydrogen nucleus PROTON

Deuterium nucleus PROTON + NEUTRON

Helium 3 nucleus PROTON + NEUTRON + PROTON

Helium 4 nucleus PROTON + NEUTRON + PROTON + NEUTR

46

How sunlight is made

This diagram shows the steps of the three-stage energy-producing process which goes on inside the sun. When the energy which is produced at each stage of the process reaches the photosphere it is radiated into space as sunlight.

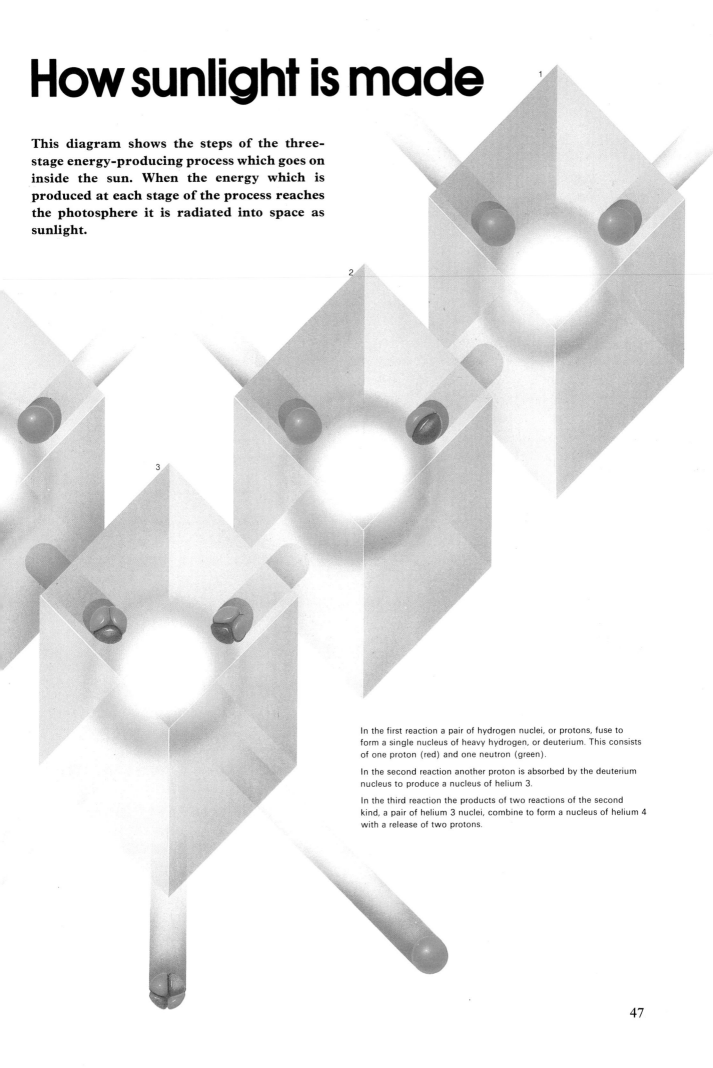

In the first reaction a pair of hydrogen nuclei, or protons, fuse to form a single nucleus of heavy hydrogen, or deuterium. This consists of one proton (red) and one neutron (green).

In the second reaction another proton is absorbed by the deuterium nucleus to produce a nucleus of helium 3.

In the third reaction the products of two reactions of the second kind, a pair of helium 3 nuclei, combine to form a nucleus of helium 4 with a release of two protons.

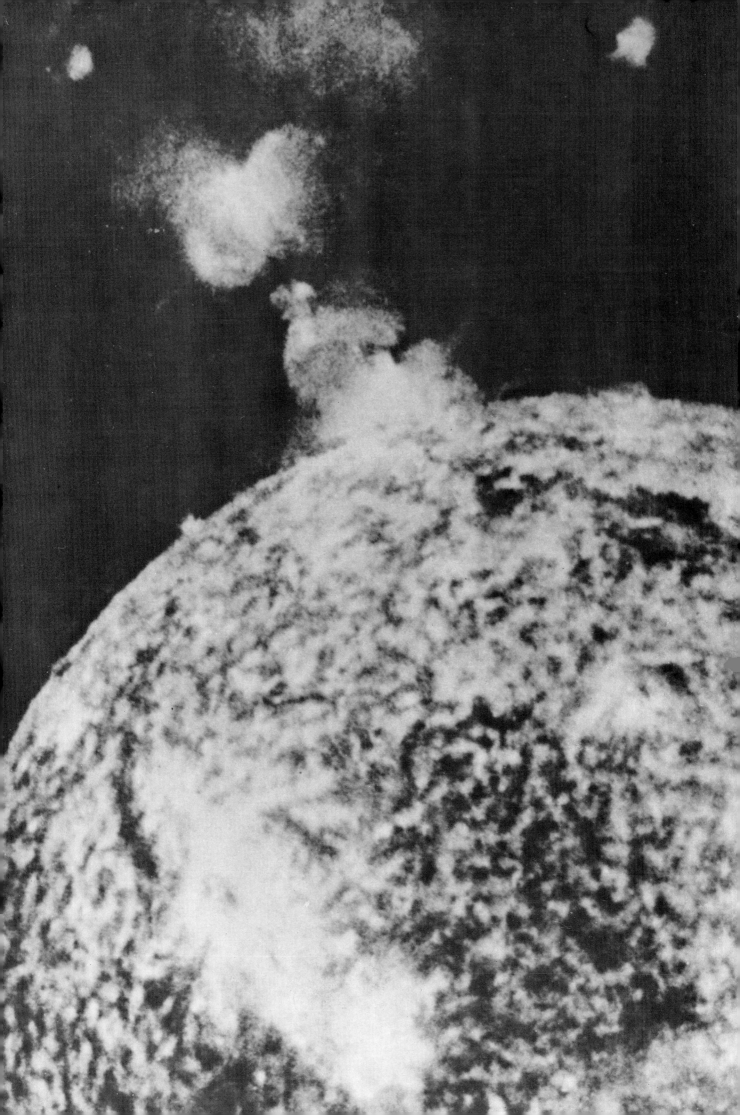

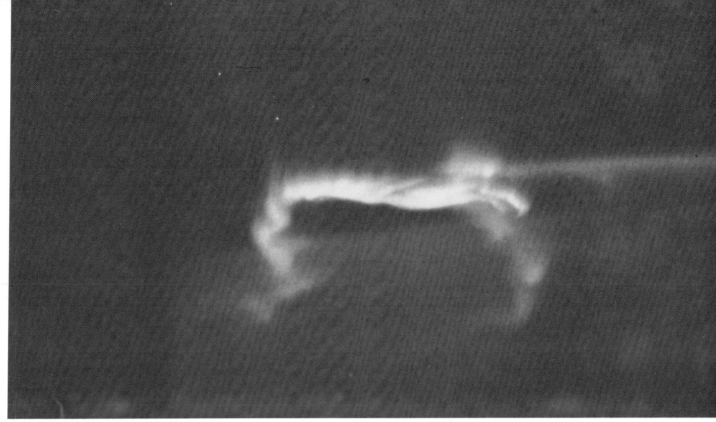

As well as giving out heat and light from the photosphere, the sun also radiates a continuous stream of atomic particles into interplanetary space through the corona. This continuous expansion of the sun's upper atmosphere is called the solar wind. Normally a relatively gentle (350 to 800 kilometres a second) breeze of hydrogen and helium nuclei, the solar wind may reach gale force at peak seasons in the eleven-year sunspot cycle.

Propelled by the sudden release of energy from flares in the chromosphere above active zones, storm clouds of hydrogen and helium nuclei (photograph, left) and other emissions gust forth through the corona at up to 2,000 kilometres a second. Because the sun is rotat-

ing, this high-speed jet of particles leaves the sun to spiral across space like water from the nozzle of a lawn sprinkler.

When, after a journey lasting several hours, the spiral intersects the earth's orbit, interactions between the energetic solar nuclei and particles in the earth's upper atmosphere produce widespread electrical disturbances, communication blackouts, compass failures and nocturnal 'lights', or aurorae.

The colour photographs, taken during the Skylab 3 mission in 1973, show aurorae in the southern hemisphere. The permanent aurora above the South Pole appears as a thin line above the edge of the earth in the background.

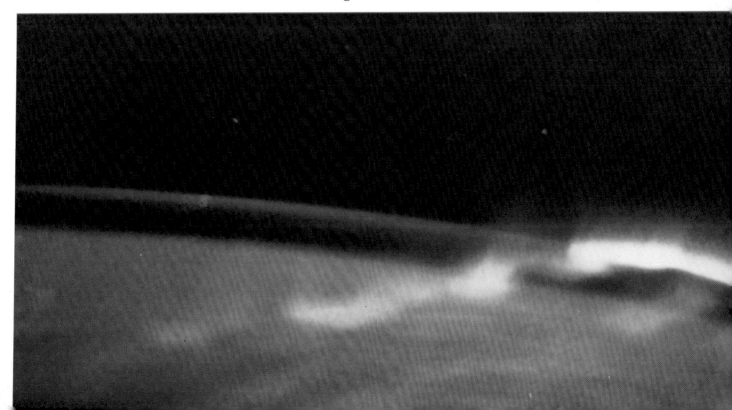

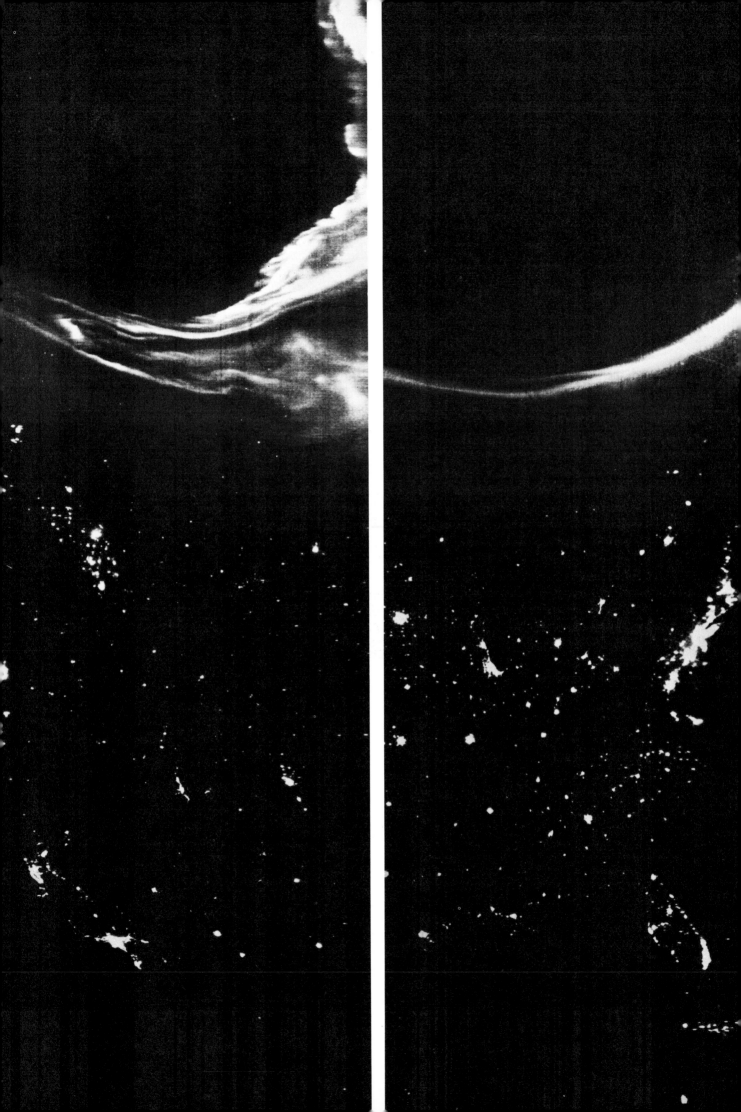

again produced by a third reaction when the components of the two helium 3 nuclei are re-combined to form a nucleus of ordinary helium, made up of two protons and two neutrons. The two spare protons are then free to fuse with others to form more helium nuclei, and so the chain continues.

Another product of these energy-producing nuclear reactions is a steady stream of sub-atomic particles called neutrinos. Travelling at the speed of light, they journey from the solar interior to the earth's surface in just over eight minutes. When a solar neutrino meets a terrestrial chlorine atom it reacts with its nucleus to produce that of a rare gas called argon. A neutrino astronomer's basic equipment is thus simple – a pool full of carbon tetrachloride, a chemical whose molecules contain four chlorine atoms. Any increase in the number of argon atoms is a measure of what is happening inside the sun. The first long-term neutrino-counting experiment was begun in 1968 by an American chemist Raymond Davis. To shield his tank of tetrachloride from other less energetic particles arriving from the sun and other sources, he placed it down the deepest mine he could find, the Homestake gold-mine in South Dakota. By pumping the fluid through some very sophisticated plumbing he was able to count the number of freshly formed argon atoms and therefore the number of solar neutrinos reaching his tank. His first results were disturbing. The number of captured neutrinos was surprisingly low. From calculations involving the internal temperature of the sun, for so long thought to be in excess of 15,000,000 °K, he was expecting more. Was the centre of the sun not so hot after all, or was his experiment wrong? Raymond Davis checked his equipment but the count remained low. Today astronomers are still discussing the question after more than eight years of solar neutrino records. About the only consensus is that the centre of the sun is much less well understood than it was once thought to be.

An even more recent and unexpected discovery was made in 1974. Another American astronomer, Henry Hill, was making very accurate measurements of the sun's diameter when he found a periodic variation of several kilometres. The sun is breathing in and out like a giant balloon. Similar up-and-down movements have since been discovered in the solar atmosphere by a team of French experimenters. Using sensors aboard an unmanned American Orbiting Solar Observatory spacecraft, they have detected a fourteen-minute, 1,300-kilometre expansion and contraction of the sun's gaseous envelope. The mechanism of these oscillations is still unknown but it may, in time, be possible to use these vibrations to study the sun's interior, in much the same way as seismic vibrations have enabled geologists and geophysicists to elucidate the inner structures of the earth and the moon.

Every second, some five million tonnes of solar fuel, hydrogen, is converted into energy. Hour by hour the sun contracts and expands, a vast globe of incandescent gas ruled by deep and mysterious magnetic forces. Day by day its surface changes, now freckled with sunspots, now strangely featureless. For billions of years the nucleus of our solar system has dispensed its life-giving energy across interplanetary space, etching its history in as yet unravelled codes in the rocky archives of the inner planets, and in the rings of trees and layers of sediments washed from ancient landscapes on earth. And although five million tonnes of hydrogen is only the minutest fraction of the sun's total mass – which only diminishes by some 150 trillion tonnes in the course of a year – after a few billion years, when much of the solar hydrogen will have been converted into helium, nuclear reactions involving heavier elements will take over. The sun will grow brighter and bigger, engulfing the orbits of the inner planets out to Mars. In time, this giant red sun will burn itself out, shrink again to about one per cent of its present size and turn white. After a life of some eleven billion years, the sun will fade and a charred cinder will occupy the centre of our dead solar system. Or will it? A few billion years is a long time, long enough for our descendants to have mastered the art of small-star maintenance.

Recorded on consecutive orbits in January 1973, the 'Northern Lights', or Aurora Borealis, is seen spanning the upper atmosphere above central Canada in this remarkable pair of night-time photographs taken by a United States Air Force weather satellite. South of the aurora, city lights outline the geography of the United States.

MERCURY

Mercury, the sun's closest neighbour, is the solar system's smallest planet. Its density, like that of the earth, is about five and a half times that of water, while its surface gravity, which is a little over a third of that of the earth, is about the same as that of Mars. Four fifths of the planet's mass is concentrated in a metallic core about the same size as the earth's moon. This is surrounded by a roughly 600-kilometre-thick mantle of rock with a heavily cratered surface.

Like all the planets, Mercury was probably formed by a gradual accumulation of material from the innermost of the nebulous rings of left-over debris which surrounded the then recently formed sun. The internal temperature of the planet, raised and maintained by numerous collisions with accreting debris, became sufficiently high for a heavy iron-rich core to develop and separate from a silicate-rich mantle. At this stage, Mercury's surface was likely to

---

Sunlight falls on a segment of the western hemisphere of Mercury, innermost of the sun's nine planets. This picture is a mosaic assembled from a number of photographs taken by the American spacecraft Mariner 10 as it approached the planet for the first time in March 1974. Until then the features of the planet's surface could only be guessed at (compare telescope image overleaf). The Mariner transmitted detailed pictures of about a third of the planet, revealing a moon-like world of craters and lava-covered plains. Like lunar and Martian impact scars, craters on Mercury vary in age. The newest ones, bright and sharp-rimmed, contrast with the battered walls of darker, older basins.

have been entirely molten and relatively smooth, although it cooled and hardened quickly enough to preserve some large early impacts which are still visible. As it cooled, the planet contracted and its surface became wrinkled like a dried plum. Where it has not been obliterated by subsequent cratering or volcanic activity, this ancient surface can still be seen today.

Looking at Mercury through a telescope can be very frustrating. Because its orbit lies inside that of the earth, the planet's back is turned when it is closest to us. When the day-time side does face us, it is behind the sun. Even at intermediate distances, when the planet is partially lit, the view is of little more than what could easily be mistaken for a used tennis ball. Our knowledge of Mercury was therefore increased more than a thousandfold by the success of the Mariner 10 mission in 1974. Using the gravitational field of Venus to deflect its path towards the orbit of Mercury, the American Mariner 10 spacecraft reached the planet in March of that year. The Mariner's orbit was now such that it returned to make a second fly-by in September and a third in March 1975. Photographs of about a third of the planet's surface were transmitted to earth.

The most spectacular feature, first revealed by the spacecraft's cameras, is a 1,300-kilometre-diameter impact scar called the Caloris Basin. The inner, flat part of the basin is ringed like a target with rifle-range concentric ridges and cracks. Its rim is made up of a ring of mountains up to 2,000 metres high. Formed by collision with a body of asteroid dimensions, the basin's margins are dotted with secondary craters caused by ejection of material following the impact.

53

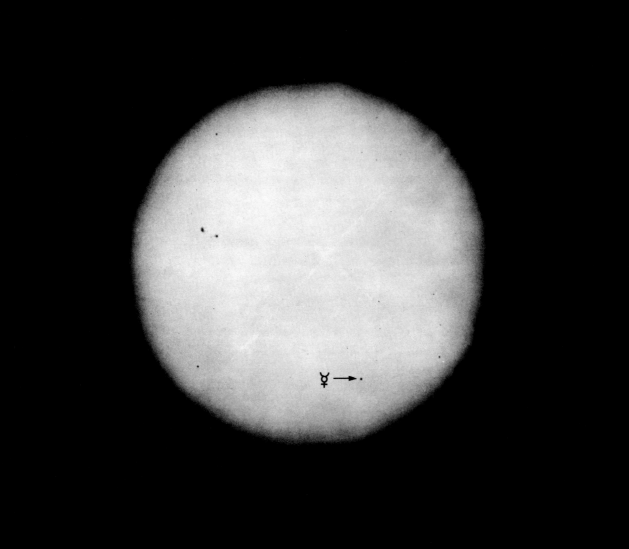

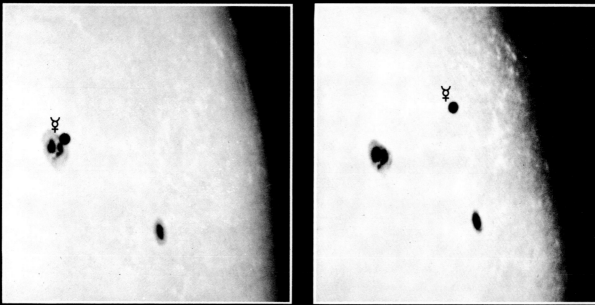

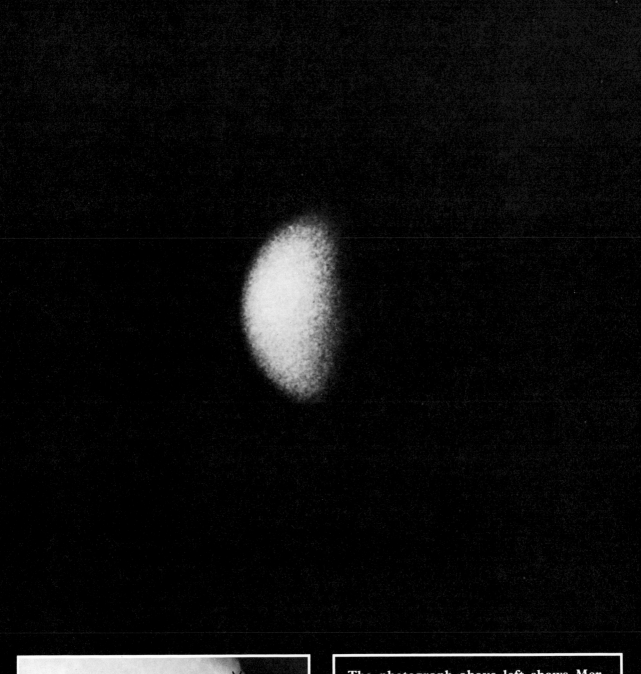

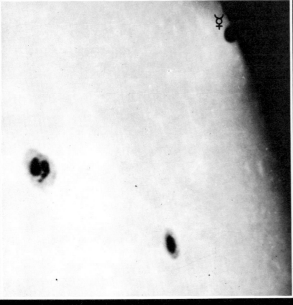

The photograph above left shows Mercury passing in front of the sun during the transit of November 1914. The sequence of three photographs below was taken during the transit of May 1970.

The photograph above right shows what Mercury looks like when viewed through an earthbound telescope.

The painting overleaf shows morning sunlight on a peak in the Caloris Montes, the circular mountain chain which surrounds the Caloris Basin, one of Mercury's most prominent features. In the foreground, a small fresh impact crater is surrounded by rays of light-coloured ejected material.

Mercury West One

This mosaic of Mariner 10 photographs shows a largely southern section of the western hemisphere of Mercury. The map on the right is a geological interpretation of the same region. A grid indicates its coordinates. Each colour corresponds to a particular kind of terrain (see key, bottom right). The small bright-rayed crater (dark red on the map) embedded in the north-western rim of an older but still relatively recent crater (bright red on the map) in the upper middle of the mosaic has been named Kuiper, after the noted American astronomer Gerard Kuiper who was a member of the team which designed and guided the spacecraft. Although it is quite small (diameter 41 kilometres) Kuiper is one of the planet's most prominent features. It can even be seen (as a bright patch on the generally dark Mercurian surface) in photographs taken while the spacecraft was still over four million kilometres from the planet. Photographs on the following pages include closer views of Kuiper and some of the other features of this region.

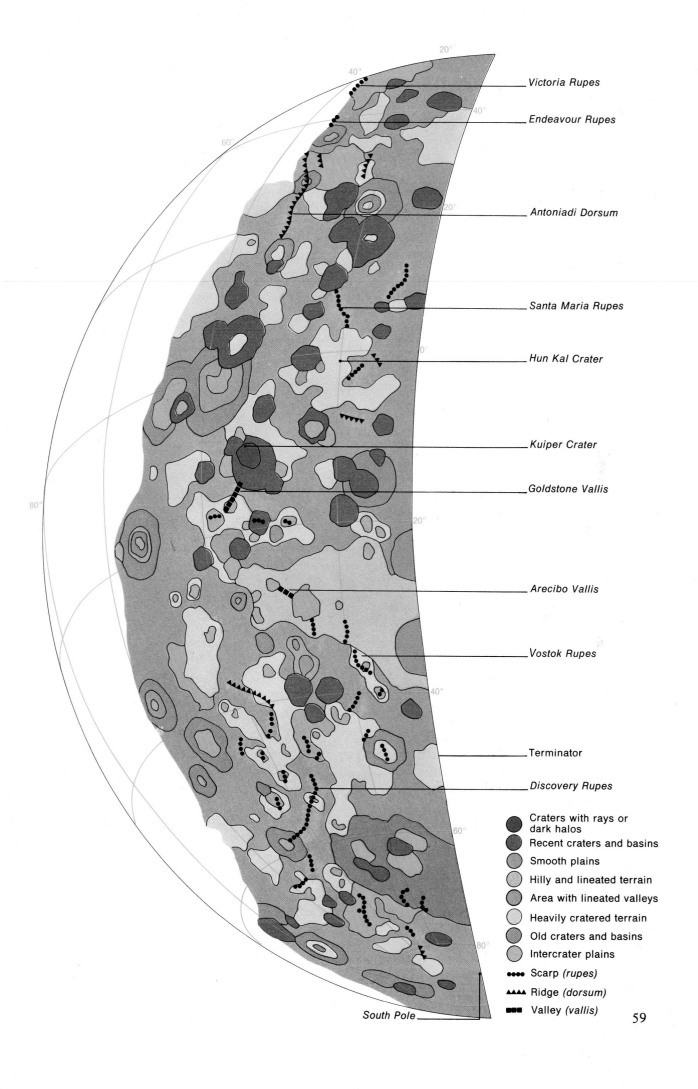

Victoria Rupes

Endeavour Rupes

Antoniadi Dorsum

Santa Maria Rupes

Hun Kal Crater

Kuiper Crater

Goldstone Vallis

Arecibo Vallis

Vostok Rupes

Terminator

Discovery Rupes

20°
40°
40°
60°
20°
20°
40°
60°
80°
80°

South Pole

Craters with rays or
dark halos

Recent craters and basins

Smooth plains

Hilly and lineated terrain

Area with lineated valleys

Heavily cratered terrain

Old craters and basins

Intercrater plains

●●●● Scarp (rupes)

▲▲▲▲ Ridge (dorsum)

■■■ Valley (vallis)

59

Arecibo Vallis, a prominent Mercurian
valley, runs north-west from the rim of
the smooth, lava-filled basin of the large
crater in the middle of this photomosaic,
centred on latitude 30° south, longitude
$26\frac{1}{2}°$. To the south-east of the large crater,
sunlight falls on the west-facing flank of
Vostok Rupes, a scarp which cuts across
two smaller basins. Several other linear
features can be seen in this generally
hilly region on the side of the planet
opposite the Caloris Basin impact zone.

These are noticeably closer to the primary impact area than are similar features on the earth's moon where lower surface gravity has allowed them a wider distribution. On the opposite side of Mercury, a hilly area appears to have been formed by transmission and focusing of the seismic energy of the impact shock through the body of the planet.

Some time after the period of heavy bombardment which gave rise to the Caloris Basin and numerous smaller craters, parts of the Mercurian surface were affected by volcanic activity. Extensive plains were formed by the lava, similar to the 'seas' on the earth's moon, though generally smaller in area in the hemisphere photographed by Mariner 10. The craters formed from then on and until the present time are often surrounded by rays of lighter-coloured material.

Mercury's rotation period – the time it takes to make one turn on its axis – is 58·65 earth days. The length of its year – the time it takes to complete a circuit of the sun – is 87·97 earth days. Thus the Mercurian year is exactly one and a half times longer than the rotation period. This relationship, which astronomers call spin–orbit coupling, affects Mercury's seasons. Unlike the earth's axis, which is tilted – so that its northern and southern hemispheres take it in turn to face the sun during their respective summers – that of Mercury's is nearly perpendicular to the plane of its orbit. Thus Mercurian seasons do not alternate according to latitude but – because of the simple relationship between the rotation period and the length of the year – according to longitude. The Caloris, or 'hot', Basin, for example, earns its name from its position near the middle of one of Mercury's relatively hotter 'summer' zones, around the 180° longitude meridian. At alternate perihelions – when Mercury is closest to the sun – the Caloris Basin experiences an extremely hot 'summer'. By the time the next perihelion comes around – a Mercurian year later – the planet has made exactly one and a half turns on its axis. It is then the turn of the hilly area on the opposite side, around the Mercurian 0° longitude meridian, to face the sun. After another year – and another one and a half turns of the planet's axis – the Caloris Basin is once again on the facing side. Mercury's orbit is very elliptical. At perihelion

Victoria Rupes, a scarp located between latitudes 45° and 55° in the northern hemisphere, can be seen above running south from the horizon for several hundred kilometres. Many of the craters in this photograph have central peaks.

Below, the crater Kuiper, named for the astronomer and Mariner 10 team member Gerard Kuiper, lies across the rim of a larger basin. Located at around latitude 11° south, longitude 32°, Kuiper is one of the brightest features of the Mercurian surface.

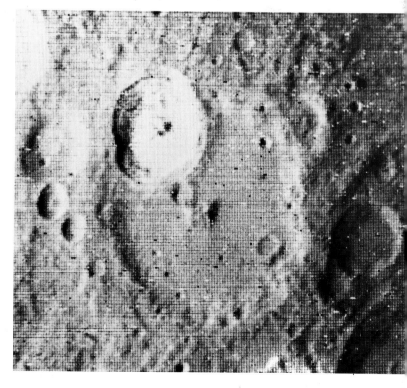

A day longer than a year

Mercury completes one turn about its axis with respect to the stars in about fifty-nine earth days while its journey around the sun lasts just under eighty-eight days. The result (see diagram, right) is a day lasting two years.

The sequence of three pictures below shows the double sunrise that would be seen by an observer at either the 90° or 270° longitude meridians when the planet is closest to the sun. The sun is seen to rise (1) but it sets again (2) as the effect of the planet's progress round the sun (shown here as shift to the right) temporarily 'cancels' the effect of the planet's rotation before rising a second time (3).

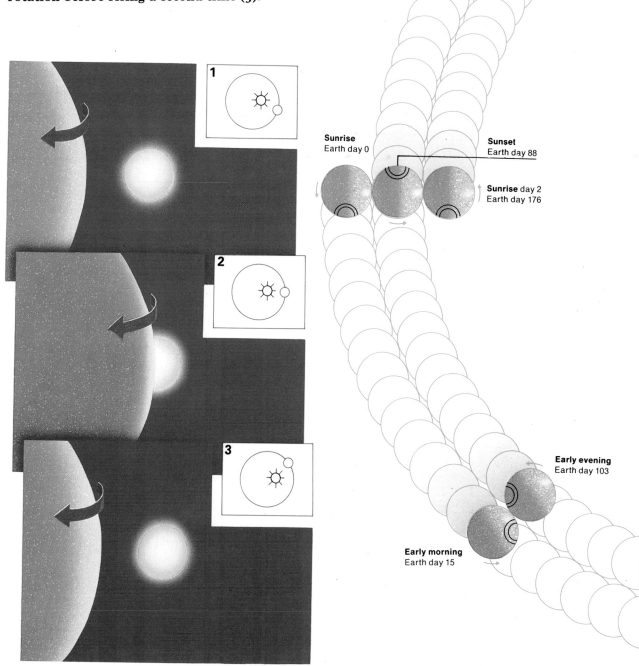

Late afternoon
Earth day 73

First light
Earth day 161

Sunrise
Earth day 0

Sunset
Earth day 88

Sunrise day 2
Earth day 176

Early evening
Earth day 103

Early morning
Earth day 15

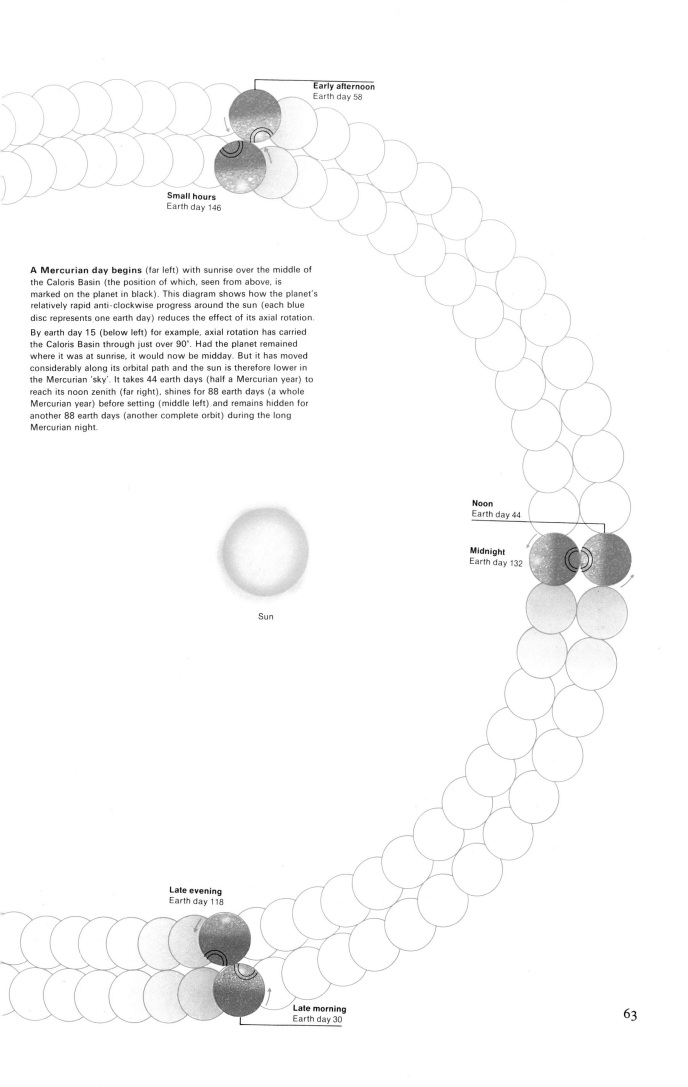

Early afternoon
Earth day 58

Small hours
Earth day 146

A Mercurian day begins (far left) with sunrise over the middle of the Caloris Basin (the position of which, seen from above, is marked on the planet in black). This diagram shows how the planet's relatively rapid anti-clockwise progress around the sun (each blue disc represents one earth day) reduces the effect of its axial rotation.

By earth day 15 (below left) for example, axial rotation has carried the Caloris Basin through just over 90°. Had the planet remained where it was at sunrise, it would now be midday. But it has moved considerably along its orbital path and the sun is therefore lower in the Mercurian 'sky'. It takes 44 earth days (half a Mercurian year) to reach its noon zenith (far right), shines for 88 earth days (a whole Mercurian year) before setting (middle left) and remains hidden for another 88 earth days (another complete orbit) during the long Mercurian night.

Sun

Noon
Earth day 44

Midnight
Earth day 132

Late evening
Earth day 118

Late morning
Earth day 30

63

the planet is around 24,000,000 kilometres closer to the sun than at aphelion, the most distant point in its orbit. Thus the monopolization of the perihelion facing positions by the 0° and 180° longitude meridians causes them to receive some two and a half times more solar energy than the intermediate 90° and 270° 'winter' meridians, which face the sun at aphelion.

Because Mercury's axial rotation is so slow, the consequent apparent east–west motion of the sun across the Mercurian sky is considerably modified by an opposing west–east motion due to the planet's movement along its orbit. Thus a Mercurian day – 176 earth days, from sunrise to sunrise – is longer than the rotation period by a factor of three, and twice as long as a Mercurian year. On earth, our relatively short day is only about four minutes longer than our planet's rotation period. At one point along Mercury's orbit, during perihelion, this day-lengthening effect becomes somewhat bizarre. When the Caloris Basin is about to face the sun, an observer watching the eastern horizon from around the 270° longitude would see the sun rise. But it would then stop in its tracks, change its mind, and set before rising again to start the long Mercurian day. This curious behaviour is a result of the temporary cancellation and reversal of the apparent east–west motion of the sun – due to axial rotation – by the apparent west–east motion – due to the planet's movement along its orbit.

At midnight on Mercury it is very cold. Temperatures fall to around 170 °C below zero. Because the planet has no atmosphere, heat accumulated by the rocky surface during the day-time leaks away into space soon after sunset. Early on in the Mercurian morning, the temperature climbs back up to over 429 °C. In three visits, Mariner 10 photographed about a third of the Mercurian surface. It remains to be seen whether the rest of the planet is similar, or whether a fourth visit by another spacecraft will reveal some surprising new features.

Just south of the equator, the comparatively tiny crater Hun Kal (arrowed) defines the 20° longitude meridian for maps of Mercury. North of Hun Kal the larger crater in the middle of this photograph measures twelve kilometres from rim to rim.

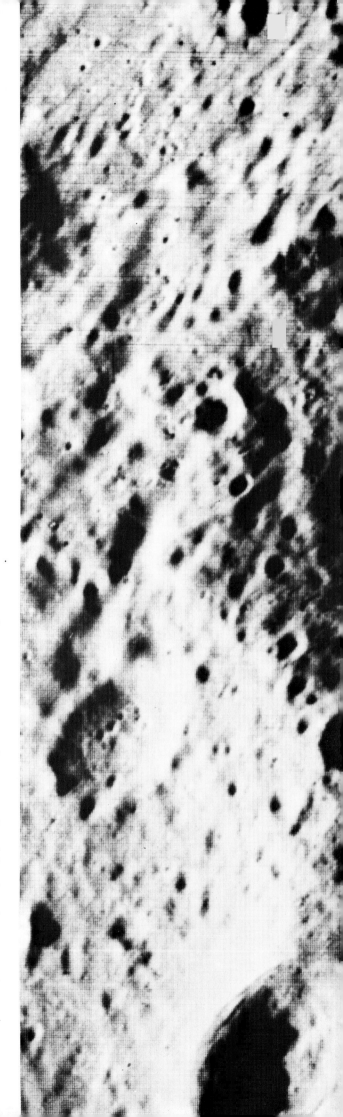

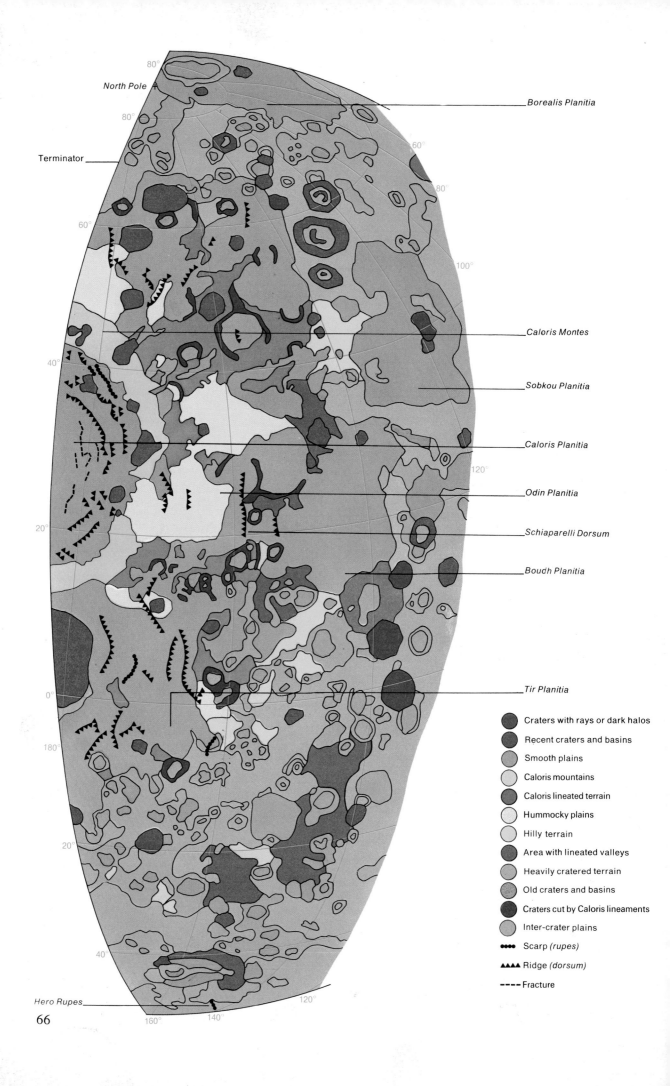

North Pole

80°
80°
60°
80°
60°
100°
40°
40°
120°
20°
20°
180°
0°
0°
20°
20°
40°
120°
160° 140°

Terminator

Borealis Planitia

Caloris Montes

Sobkou Planitia

Caloris Planitia

Odin Planitia

Schiaparelli Dorsum

Boudh Planitia

Tir Planitia

Hero Rupes

66

Craters with rays or dark halos

Recent craters and basins

Smooth plains

Caloris mountains

Caloris lineated terrain

Hummocky plains

Hilly terrain

Area with lineated valleys

Heavily cratered terrain

Old craters and basins

Craters cut by Caloris lineaments

Inter-crater plains

●●●● Scarp (rupes)

▲▲▲▲ Ridge (dorsum)

---- Fracture

Mercury
West Two

This mosaic of Mariner 10 photographs shows a largely northern section of the western hemisphere of Mercury bordering its western junction with the eastern hemisphere (180° longitude). The prominent feature of this region (centre of left edge) is the eastern half of Caloris Planitia, an enormous basin surrounded by the 2,000-metre-high peaks of Caloris Montes. A closer view of these mountains (apple green in the geological interpretation on the left) can be seen on page 69. With a diameter of 1,300 kilometres, the Caloris Basin, as this ringed structure is also called, is comparable in scale to the Imbrium Basin on the earth's moon.

South-east of the Caloris Basin, Tir Planitia is one of several plains in this sector. Tir was probably formed when lava floods, associated with the event which formed the Caloris Basin, covered an older rougher surface. Closer views of parts of Tir and Borealis Planitia and the scarp Hero Rupes can be found on the following pages.

The area in the photograph above lies between latitudes 70° and 85° north and longitudes 90° and 150°. A heavily cratered surface in the foreground forms the southern margin of the comparatively smooth Borealis Planitia, or northern plain, which can be seen in the background.

Below, a bright area, south-east of the Caloris Basin in the north-eastern corner of Tir Planitia, where an ancient heavily cratered surface has been partly overlain by lava flows to produce smooth areas between the crater rims. The small young crater in the centre of the light patch is ten kilometres across and is located at latitude 12° north, longitude 170°.

The left-hand half of the mosaic on the right shows the mountainous eastern rim of the huge Caloris Basin. Centred on latitude 30° north, longitude 190°, it is the largest known Mercurian feature. Like the Imbrium Basin on the earth's moon, it appears to have been formed by an impacting body of asteroid dimensions. The event produced vast bodies of molten rock which erupted on to the surface to form extensive plains.

The mosaic above, composed of over 200 separate photographs taken by Mariner 10 during its second fly-by of Mercury in September 1974, shows the planet as seen from a spacecraft passing beneath it. The south pole (arrowed) is in the middle of the terminator (day–night boundary) at the bottom of the mosaic. Note numerous prominent rayed craters. The crater Chao Meng-Fu which contains the pole is around 170 kilometres across.

The photograph on the left is a detail from the mosaic showing an area located about midway between the south pole and the small bright rayed crater near the left edge of the planet. The centre of this region is crossed by a cliff, Hero Rupes, more than 300 kilometres long, which may have been formed by crustal shrinkage. The zone encompassing this cliff and the bright crater can also be seen in another detail overleaf.

A detail from the mosaic shown on the previous page depicts a heavily cratered zone in the southern hemisphere of Mercury. The bright rayed crater on the left is located around latitude 32° south, longitude 172°. The south pole is beyond the terminator to the right of the picture.

The painting overleaf shows the sun rising over this harsh landscape where the baked edge of a lava flow (left) encroaches across a cratered plain.

VENUS

An earth twin in terms of size, Venus is also our closest planetary companion. Moving around the sun at a swifter pace than we do, it passes within 40,000,000 kilometres of the earth every nineteen months. These encounters are more frequent, and on average twice as close as those with our other, outer neighbour, Mars. But because its earth-facing hemisphere is then in darkness and its surface lies permanently hidden beneath an unbroken blanket of cloud, Venus is by far the lesser known of the two. Much, until recently, was left to the imagination. Uncluttered by facts, astronomers were free to paint all manner of science-fiction land- and seascapes waiting to be discovered beneath the Venusian smog.

In 1879, the American astronomer Percival Lowell, already famous for his maps of Martian canals, published a chart showing a similar network of markings on the surface of Venus. In 1918, the Swedish chemist and Nobel prize-winner, Svante Arrhenius, envisaged a world of

A close-up view of the outer veils of Venus, this photomosaic was assembled from pictures transmitted to earth by the American spacecraft Mariner 10 as it flew past the planet on its way to Mercury in February 1974. Upper atmosphere clouds of sulphuric acid and carbon dioxide crystals girdle the planet. Venusian winds blow east to west (right to left) from the equator towards the poles. Composed mainly of carbon dioxide, the Venusian atmosphere is nearly one hundred times thicker than the earth's and acts as a global 'greenhouse', trapping heat from the sun to produce surface temperatures of 475 °C.

steamy jungles teeming with primitive life. Life, it was suggested, had begun on Venus in much the same way as it had begun on earth – but for some reason the progress of Venusian evolution had only reached the Venusian equivalent of the coal-forest era which flourished on earth some 300,000,000 years ago. (This attractive image of a green and pleasant world was still popular in the 1950s when the English comic-strip hero, Dan Dare, arrived on the planet to confront his lifelong adversary, the tyrant Mekon.)

A few years ago it was realized that the transparent component of the eighty-kilometre-thick Venusian atmosphere – mostly carbon dioxide – forms a planetary sun-trap, and that the surface of Venus – like the floor of a greenhouse – must be very hot. Then the jungle idea went out of fashion. Some astronomers now pictured Venus as a planet of endless deserts, others as a sphere completely submerged by oceans. It was even suggested that the clouds were made up of droplets of oil, and that beneath them lay seas of petroleum – a good example, perhaps, of necessity being the mother of invention.

With the coming of the space age, the truth began to emerge. In December 1962 remote sensing equipment aboard Mariner 2, an American spacecraft flying within 35,000 kilometres of the planet, indicated a surface temperature of about 428 °C – higher than that of Mercury. In October 1967 and May 1969 Russian Venera spacecraft encountered correspondingly high temperatures in the greenhouse atmosphere and in December 1970 Venera 7, reporting directly from the surface, recorded a blistering 475 °C.

In the summer of 1975, the Venusian clouds

When opposite the earth, beyond the sun, Venus is small and full (above). Moving round the sun, relative to the earth (left), it grows larger but lies increasingly in shadow until closest approach (below) when it becomes a crescent.

were penetrated by radar beams from the 305-metre dish of the Arecibo radio telescope in Puerto Rico. Analysis of their echoes enabled American astronomers to construct a picture of an area in the northern hemisphere the size of Canada, extending 80° in longitude between the 46° and 75° latitude meridians (see page 83).

This region is dominated by a large pear-shaped basin the size of Hudson Bay. Except for two small bright areas, which could be impact craters, the basin floor appears dark. In a radar image of this kind, darkness may mean smoothness, and smoothness, in this case, may be due to lava flows. The darkness of the basin floor contrasts strikingly with its lighter, and therefore rougher, surround, which is thought to be a series of mountainous ridges. Because the ridges which surround 'Hudson Bay' do not form a circle – unlike those which border major craters on Mercury and the earth's moon – they may be the result, not of an asteroid impact, but of mountain-building processes – gradual rock movements, similar perhaps to those which occur on earth. On the other hand, to the southwest of the basin, a lighter (and therefore rougher) area resembles those regions on the earth's moon which are covered with debris thrown out when large objects hit the lunar surface. Refinement of another set of radar data, also obtained in 1975, has led to the suggestion that there may be volcanoes on the planet.

The Venusian atmosphere is extremely dense. On Venus, the surface pressure is ninety times greater than it is on earth. The clouds are pale yellow. They are thought to be made up of droplets of concentrated sulphuric acid – their colour may be due to the presence of elemental sulphur and/or one or more of its compounds. Worse still, from the spacecraft designer's point of view, the fluorine content of the atmosphere may be combined with the sulphuric acid to form fluorosulphuric acid – the strongest of the simple mineral acids. Venusian rain may well be the most corrosive fluid in the solar system, and this may explain why – even though they are designed to survive temperatures greater than that of molten zinc and pressures similar to those encountered at depths of 550 fathoms in the earth's oceans – spacecraft which have landed on Venus do not remain in working order for very long.

Such was the fate of a pair of Russian Veneras, 9 and 10, which arrived on the planet in October 1975. Venera 9's transmissions ceased after only fifty-three minutes. Those of Venera 10 lasted twelve minutes longer. The data returned to earth included landing-site temperatures of 485 °C and 465 °C, and pressures of 90 atmospheres and 92 atmospheres. They also transmitted the first pictures of the Venusian landscape – two panoramas showing dry rocky surfaces. Visibility was better than expected. Both spacecraft were equipped with lights but the sky turned out to be as bright as it is 'on a cloudy day in Moscow in June'. With the high temperatures and corrosive nature of the Venusian atmosphere in mind, astronomers expected to see an erosion-polished surface. They were therefore somewhat surprised by the sharp outlines and rough edges of some of the boulders in the Venera 9 photograph (see pages 86–7). The presence of unweathered rocks is an indication of either a relatively young landscape, or a geologically old landscape little modified since its formation. It is possible, however, that on Venus, as on earth, geological forces are at work. There are also further indications of volcanic activity in the shelf-like appearance of the Venera 10 landing site which may be the weathered surface of a lava flow. Further Russian Venera missions and the American Pioneer spacecraft due to map Venus by radar from orbit in 1978 will tell us more about the geological forces which shape the planet's hidden landscapes.

Were it visible at the Venusian surface, the sun would rise in the west and set in the east because, seen from above, Venus rotates in a clockwise direction, unlike most of the other planets and moons, whose rotation is anti-clockwise. It rotates very slowly – its retrograde rotation period is 243 earth days. A Venusian year – one complete journey round the sun – takes 225 earth days. This combination gives the planet a day – from sunrise to sunrise – 117 times longer than an earth day. It also means that the same

The painting overleaf shows billowing clouds of sulphuric acid droplets building up over a mountainous region near the Venusian equator. Their yellow colour is due to the presence of sulphur.

79

Venusian hemisphere is facing the earth when the two planets are closest every nineteen earth months, or five Venusian days.

In photographs taken by the American spacecraft Mariner 10 – which flew within 5,800 kilometres of the planet on its way to Mercury in February 1974 – Venus looks a little like the earth. Bands of lighter-coloured cloud stand out against a dark background. But the similarity is deceptive. The Mariner photographs were taken in ultraviolet light – the darker areas are not glimpses of the surface but possibly upper atmosphere cloud regions containing a relatively higher proportion of sulphuric acid droplets.

Scudding from east to west, the clouds spiral outwards from the equator towards the poles. Wind speeds of up to 360 kilometres an hour have been recorded by Russian Veneras and, by tracking its formations around the globe, astronomers have found the upper atmosphere rotation period to be about four earth days. Spacecraft measurements indicate calmer conditions in the lower part of the atmosphere and at the surface.

Windswept or not, life on the Venusian surface today is clearly unthinkable. The same probably holds true for the lower half of the atmosphere. At an altitude of forty-five kilometres the temperature is still higher than the boiling point of water but at fifty-five kilometres it is close to freezing point. The pressures accompanying this ten-kilometre earth-like temperature spectrum vary between two atmospheres at the base and a quarter of an atmosphere at the top. Except for the composition of the atmosphere – more than 95 per cent carbon dioxide with less than one per cent water – conditions at an altitude of fifty kilometres above the Venusian surface are not unlike those found at tropical latitudes on earth. And in this environment on earth there prospers a vast photochemical industry – the extraction of carbon dioxide from, and the release of oxygen into, the atmosphere – the industry of plant life. Until plants were evolved – more than two and a half billion years ago – the earth too had an atmosphere rich in carbon dioxide. Without plants, its present atmosphere might well become similar to that of Venus. With them, might that of Venus become more like the earth's?

The art of making other planets habitable is

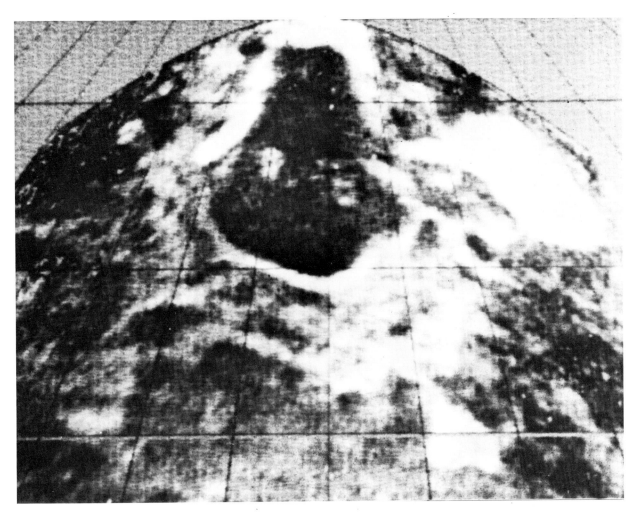

Taken seven hours apart, the Mariner 10 photographs which went to make up the three mosaic images of the planet on the left show the high-speed rotation of the Venusian upper atmosphere (note movement of arrowed cloud formation). While the body of the planet takes 243 earth days to complete a clockwise turn about its axis, the cloud tops are wind-driven around the globe in the same direction in just four earth days. There is also a tendency for clouds to move outwards from the equator towards the poles.

The radar image above, centred on longitude 325°, latitude 65° north, shows a pear-shaped area the size of Hudson Bay. Darkness, in such pictures, may denote a smooth surface, which might be that of a lava flow similar to those which have created vast plains on the faces of the other inner planets and the earth's moon. The lighter, therefore rougher, margins of this apparent basin are perhaps mountainous. The region immediately south of the basin has a radar profile similar to that of zones around large lunar craters where coarse material, ejected by impacts, blankets the surface. A rough patch, shaped like a bird's

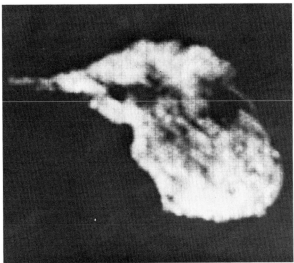

head, to the east of the basin (lower contrast detail, above), shows a variety of surface texture.

It is not yet known whether rougher zones are produced by impacts or Venusian mountain-building processes. But Venus, like the earth, has a warm interior. The painting overleaf shows a body of lava erupting on to the Venusian surface beneath a dark and corrosive vapour-filled sky.

83

Venus unveiled. Transmitted back to earth in October 1975 by two Russian Venera spacecraft a few minutes before they were overwhelmed by the hot crushing and corrosive Venusian atmosphere, these two panoramas ended centuries of speculation by showing what the surface really looks like. Venera 9 landed on a stone-covered slope (above). The rocks near the base of the spacecraft in the foreground are up to 70 cm long and 20 cm high. The photograph from Venera 10 (below) shows the spacecraft resting on an apparently horizontal rocky shelf which is slightly raised above its darker surroundings. Similar shelves of lighter coloured rock can be seen in the background.

The slope of the Venera 9 landing site and the presence of loose angular rocks suggests that Venus is a geologically active planet. On earth such features are produced when the ground is raised by mountain building and then eroded by wind, rain and running water. On Venus the ground may be similarly raised to be cracked by heat and eroded by the planet's corrosive atmosphere. It has been suggested that the smoother surface of the Venera 10 landing site is due to chemical weathering of a lava flow and that the irregular dark patches reflect local differences in its composition where the rate of erosion has been more rapid.

Moving in a faster orbit than ours, Venus

passes between the earth and the sun every nineteen months. Such Venus-earth close encounters are due in November 1978, June 1980, January 1982, August 1983, April 1985, November 1986, June 1988, January 1990 and so on. Venus-bound spacecraft are often launched a few months before, to arrive a few months after, these dates. The Americans were planning to send a Pioneer orbiter and atmospheric probes to arrive in December 1978. For landers we must look to Russian Venera missions.

The painting overleaf shows sunlight reaching the Venusian surface through a gap in the clouds. Such a gap might be produced artificially by sowing the atmosphere with algae.

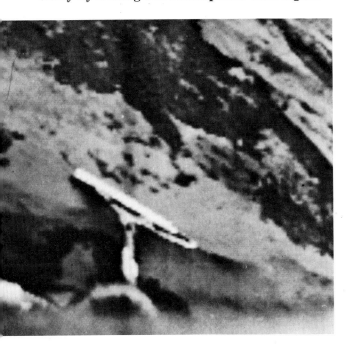

called 'terraforming'. Most projects of this nature envisage engineering on a truly cosmic scale, the costs of which would be literally astronomical. By comparison, terraforming Venus, it has been suggested, might be accomplished for next to nothing. In 1961, the American physicist and planetologist Carl Sagan pointed out that a substantial reduction of the Venusian atmosphere's carbon dioxide content would result in a correspondingly substantial lowering of its temperature and pressure. By releasing its oxygen content, the unwanted carbon dioxide could be used to convert the Venusian environment into one as habitable as the earth's. The necessary chemical engineering could be done by plants.

Cyanidium caldarium is the botanical name of one of many species of blue-green algae. A single-celled organism, its preferred earthly habitat is a hot-water spring. Fond of carbon dioxide, it reproduces itself by simple division – a single cell becomes two, four, eight, sixteen, thirty-two, sixty-four, etc. – which means that, in the right circumstances, it multiplies with increasing rapidity. In a laboratory test in 1970, the alga was introduced into a simulated Venusian atmosphere – carbon dioxide, a little acidic water, high temperature and pressure. It thrived. As the cells began to divide, the carbon dioxide thinned out and oxygen began to build up in the laboratory flasks. Many more experiments will be needed before algae are set to work on the Venusian atmosphere. Its exact composition will have to be known and tested to ensure the selection of the most suitable algal strains.

The scenario goes something like this. Consuming ever increasing amounts of carbon dioxide, the algae considerably reduce the Venusian atmosphere's capacity to retain heat. As the 'greenhouse effect' wears off, the upper atmosphere cools down and water droplets are formed – calculations suggest that there is enough atmospheric water to soak the entire planet with about 250 centimetres of rain. The first rains will not reach the ground. It will still be too hot. But they will cool the lower atmosphere, paving the way for the oxygenating algae. Eventually the rains will fall on the surface. Where the rocks are impervious, rivers will run and lakes will form. Through an oxygen sky, the sun will shine down on a truly new world . . .

EARTH

Four thousand six hundred million years ago the internal temperature of the young cratered earth was already quite high as a result of collisions between the numerous sun-orbiting bodies which had formed it. Now it was further raised by the decay of its radioactive elements. The planet became a globe of molten slag and a core of liquid iron separated from a silicate-rich mantle. As its surface cooled, a crust formed through which clouds of volcanic steam and hydrogen-rich vapour escaped into the primitive atmosphere.

Today the crust is discontinuous, being made up of a number of closely fitting plates like a suit of armour (see diagram overleaf). Each of these plates floats on the mantle below, which – warmed by the continuing decay of radioactive elements within it – is very slowly turned over by a series of large-scale convection currents. Thus its upper part – like the surface of any gently simmering liquid – is in motion. In some places material is being brought up from below. In others it is gradually moving across the surface or being drawn back down into the mantle.

The earth and its moon are shown to scale in this composite of Apollo photographs. That of the earth was taken by outward-bound Apollo 17 astronauts in December 1972 – midsummer in the sunwards-tilted southern hemisphere. Africa, Arabia and the Antarctic ice cap are clearly visible. That of the moon was taken by the departing Apollo 16 crew in April 1972. Although none of the far side of the moon (right side of the globe) would be visible from this angle, this is how the earth and moon – almost a double planet system – might appear from a distant point in space.

The plates are made up of two kinds of material – lighter, granitic, continental crust and heavier, basaltic, oceanic crust. Where they are composed of ocean crust, the plates may be joined by a mid-ocean ridge – a seam into which fresh basaltic rock is being slowly introduced from below, forcing the older parts of the plates apart. The gradual expansion of the basaltic ocean basin also increases the separation of its shores – the granitic, continental parts of the two plates – giving rise to continental drift. Elsewhere plate junctions are the sites of collisions. Where ocean crust meets ocean crust, collisions are resolved by material being taken back down into the mantle. Where ocean crust meets continental crust, the ocean crust is subducted to form an ocean trench and the margin of the continental crust which stays on top is compressed to form mountains.

The break-up of the earth's primitive crust into plates may have been caused by the last few impacts which made up the body of the planet. By punching holes in the ancient crust, colliding asteroids may have also been responsible for the earth's first ocean basins which were filled, more than 3,700 million years ago, when the rain began to fall. And while the elevated regions remained above water, the rains eroded and transported them – particle by particle – into the surrounding seas. But beneath the oceans the crustal plates went on growing and where plate margins came together they sometimes did so beneath vast basins of sediment brought down to the edges of the land masses by rivers. In this case the convergence of two plates crushed these enormous bodies of mud, silt and

Earth machine

The diagram below shows how the mountain-building earth machine is powered by deep convection currents (green arrows) in the earth's mantle.

Washed down by rivers, sediments build up in troughs along the edges of continents (left). Where these troughs overlie a plate boundary (right), underthrusting of one plate by another may compress the sediments. The only direction they can find room to move is up. New mountains are the result. The process weakens the crust and molten rock may rise through cracks to produce volcanoes.

The map on the right shows the main seams and colliding plate boundaries discovered so far. The Andes and Himalayas are both examples of comparatively recent collisions between plates where sediments have been kneaded into mountain ranges. The cores of the continents are made up of the remnants of ranges built by earlier collisions, the approximate ages of which are given in the key below. Through geological time, the shifting of plates alters the atlas pattern of land and sea. Earlier continental arrangements are shown overleaf.

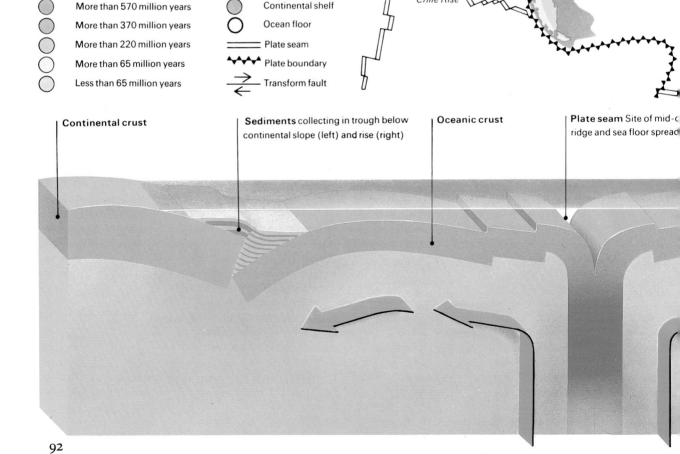

● More than 570 million years	● Continental shelf
● More than 370 million years	○ Ocean floor
● More than 220 million years	═ Plate seam
○ More than 65 million years	◆◆◆ Plate boundary
○ Less than 65 million years	⇄ Transform fault

Gorda Plate

San Andreas Fault

North American Plate

Caribbean Plate

Cocos Plate

Mid-A... Ridge

Andes

Nazca Plate

Peru-Chile Trench

South American Plate

Chile Rise

Continental crust

Sediments collecting in trough below continental slope (left) and rise (right)

Oceanic crust

Plate seam Site of mid-o... ridge and sea floor spread...

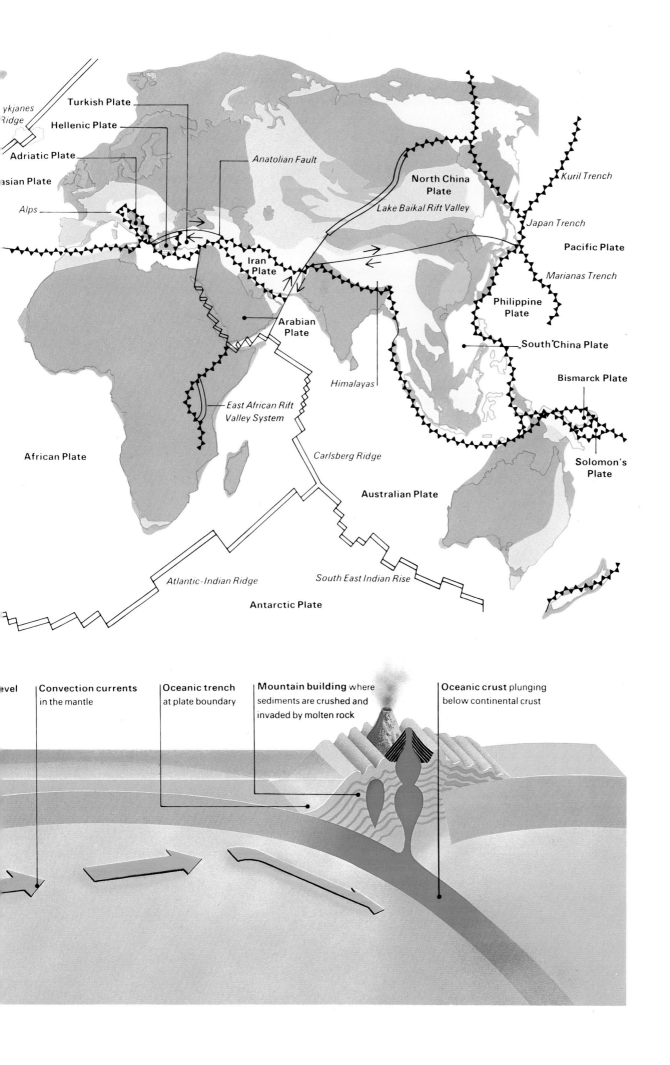

Turkish Plate

Hellenic Plate

Adriatic Plate

ykjanes
Ridge

asian Plate

Alps

Anatolian Fault

North China
Plate

Lake Baikal Rift Valley

Kuril Trench

Japan Trench

Pacific Plate

Iran
Plate

Marianas Trench

Arabian
Plate

Philippine
Plate

South China Plate

Bismarck Plate

East African Rift
Valley System

Himalayas

African Plate

Carlsberg Ridge

Solomon's
Plate

Australian Plate

Atlantic-Indian Ridge

South East Indian Rise

Antarctic Plate

evel

Convection currents
in the mantle

Oceanic trench
at plate boundary

Mountain building where
sediments are crushed and
invaded by molten rock

Oceanic crust plunging
below continental crust

Jigsaw puzzle planet

This series of globes traces the wanderings of the earth's continents – the result of several hundred million years of being pulled apart or pushed together by growing and shrinking crustal plates (see previous page). The globes are tilted to show the northern hemisphere on the left and the southern on the right. The world map for twenty-five million years ago (bottom right) was not so very different from that of the present time and may be used for comparison with the others.

Africa (1), North and South America (2 and 3), Antarctica (4), the bulk of Asia (5), Australia (6), northern Europe (7), southern Europe (8) and India (9) are shown in the positions they occupied at five different stages of geological history. For easy recognition they are shown with their present coastlines (blue outline) although the extent of dry land varied considerably with changes in sea level – the green areas represent its maximum extent, during periods when the sea was at its lowest level and the oceans retreated to the edges of the continental shelves.

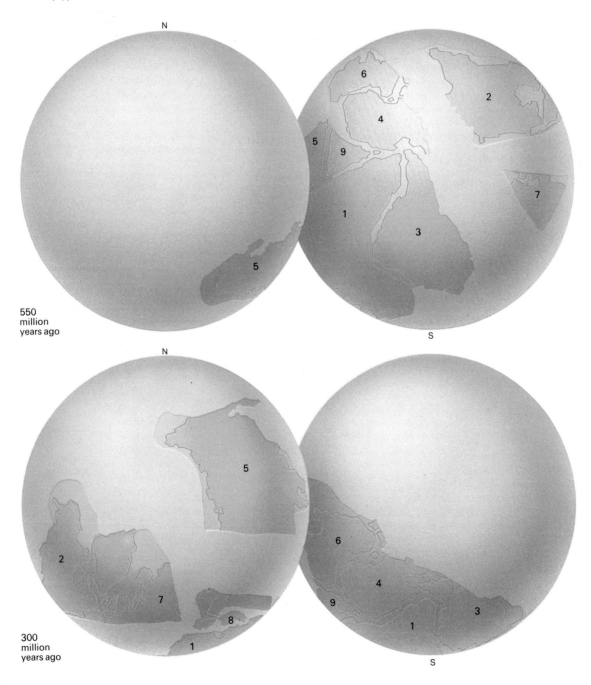

550 million years ago

300 million years ago

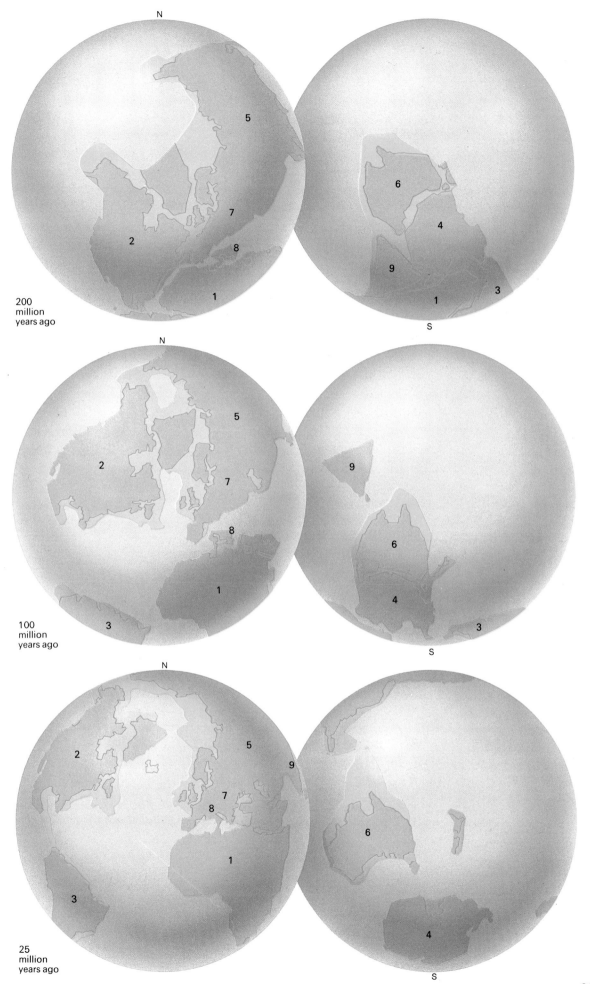

200
million
years ago

100
million
years ago

25
million
years ago

sand against the margins of the older land surfaces, folding and lifting them to form a fresh range of mountains and volcanoes – the explosive tips of large bodies of molten rock which came bubbling up through cracks in the colliding crust.

Thus the last 3,700 million years have seen a gradual accumulation of newly elevated surfaces around the cores of the older land masses, which can still be seen in Canada, Greenland, Scandinavia, eastern South America, Africa, Arabia, India, Siberia, Western Australia and Antarctica. But the present number of these continental nuclei is deceptive because many, if not all of them, were once joined together. For as they float upon the mantle, the continental pieces of the crustal jigsaw puzzle are in constant relative motion. Some 200 million years ago (see pages 94–5) the age of dinosaurs saw Asia and northern Europe joined to North America through Greenland. South America lay closer to North America than it does today. The northwestern shoulder of Africa nestled into the Gulf of Mexico while its south-eastern flank was joined to Australia through Antarctica, with India folded in beside Madagascar. The seam which gave birth to the Atlantic Ocean was barely open.

To fill the earth's ocean basins either the primitive atmosphere, volcanoes, or both these sources of water produced some 1,360 million cubic kilometres of liquid. If the planet's new blue sky then appeared through gaps in the clouds, it did so through a cooler, less dense mixture of ammonia, carbon dioxide, carbon monoxide, hydrogen, methane and nitrogen. There was no oxygen in the atmosphere, for oxygen is thought to be a consequence of, not an antecedent to, life. Without oxygen, there was no atmospheric ozone layer. And in the absence of an ozone layer, there was nothing to prevent the full blast of solar ultraviolet rays from reaching the watery surface of the planet.

Within the last fifteen years, experiments have shown that when the components of the earth's primitive atmosphere are mixed with water and exposed to ultraviolet radiation, or subjected to electrical discharges similar to those which produce lightning, then a number of more complex substances are formed. The products, though unimpressive compared with those of Count Dracula, do include amino acids – one of the fundamental molecular components of living matter.

We may therefore envisage the chemistry of life beginning on earth over 3,700 million years ago in what geologists call archaean times, at or near the surface of large bodies of water. Synthesized by the action of ultraviolet radiation or electrical discharges upon atmospheric gases, amino acids and other biologically active molecular components were formed in the earth's oceans, lakes, rivers and ponds. Shielded by their watery environment from further doses of ultraviolet (which would have had an inhibiting effect), their subsequent combination created proteins and other life-forming molecular chains. This process may, according to an American biochemist James Lawless, have been aided by clay. Certain clays, common on shorelines, are now known to favour the selection and concentration of biologically important molecules.

The spark of life may have been the first successful molecule of nucleic acid – the substance from which genes are made. Modern genes are made of *deoxyribonucleic acid* – DNA for short. The structure of DNA was unravelled in 1953 by two Nobel-prizewinning biochemists Francis Crick and James Watson. They found it to be made up of a pair of very long molecular threads wound helically around each other – the double helix. Slung between these threads – at regular intervals, like the treads of a spiral staircase – are thousands of cross-links, each of which is formed by a pair of chemical sub-units called bases. In the DNA molecule there are only two possible cross-link base pairs – adenine with thymine (A-T) and cystosine with guanine (C-G).

When the double helix unwinds the base pairs separate and the molecule 'unzips'. Along its length each thread now holds a sequence of several thousand bases – like unoccupied pegs on a washing line. Now if at one point along one thread the sequence of bases is CCATGGC, etc., then the corresponding sequence at the same position along the other thread – determined by the rule that adenine always combines with thymine and that cystosine always combines with guanine – will be GGTACCG, etc. Thus one untwisted DNA thread is the 'negative' of its partner. And just as copies can be made from photographic negatives, so the two halves of a DNA molecule can reproduce each other by matching their unpaired bases with those of fresh DNA ingredients. And this is the very secret of life; for the sequence in which the thousands of bases are arranged along the threads of DNA molecules is written in the four-letter alphabet language of the genetic code.

Not only do DNA molecules reproduce themselves – passing on their store of coded information to their offspring – they also produce templates for the mass production of all the other chemical ingredients of life. Proteins, for example, which make up 18 per cent of human body weight, consist of amino acid chains assembled according to sequences dictated by DNA molecules. Thus the evolution of nucleic acids brought order to the 'primordial soup' and paved the way for the development of the first living cells – the building blocks of biology, within each of which DNA-controlled nuclei direct the activity of millions of molecules arranged so that some of their number form a protective shield around the rest.

The first living cells were bacterium-like, similar to those whose fossils are found in the Fig Tree Chert – a 3,400-million-year-old rock formation found in Swaziland. Each time one of these simple cells divided, each daughter cell was provided with a copied set of the parent cell's complement of DNA molecules. Thus the continuity of successful cells was assured. But every now and again transcription errors would occur – part of the base sequence along a DNA molecular thread might be missing or jumbled up. The resulting offspring would then be slightly different from the parent – a mutant form would arise.

To feed themselves the bacteria drew upon the remaining reservoir of primordial soup. But this was of a limited size and must eventually have become markedly diminished. When and where this happened, natural selection would have greatly favoured those mutant cells which had 'learned' to do without it. This was made possible by the evolution and enshrinement in the DNA code book of photosynthesis – the photochemical process by which cells make their own food by reacting water with atmospheric carbon dioxide in the presence of a pigment and sunlight to produce sugars and oxygen. And so in ancient rock formations we find – alongside the fossil bacteria – traces of another kind of cell similar to those of contemporary one-celled members of the plant kingdom – the microscopic blue-green algae.

Thus, long before Prometheus, the ancestors of the blue-green algae brought fire to the planet because free oxygen is a highly reactive substance and the nature of its reaction with many other substances is combustion. And since many of the internal constituents of cells are vulnerable to oxygen, natural selection now promoted those organisms which could either expel or exploit the unwanted corrosive gas. Those which expelled it – the forebears of all modern plants – now brought about a feat of planetary engineering every bit as impressive as the shift of continents. For although it took them several hundred million years, they gradually replaced the carbon dioxide content of the earth's atmosphere with oxygen. Those which exploited it – the forebears of the animal kingdom – were able to do so by controlling the exposure of cellular fuel to oxygen and by making constructive use of the resultant combustion energy. Thus the oxygen produced by cells which had adopted the plant way of life – energy from sunlight – allowed the development of the animal way of life – energy from combustion. It is not only for their food that animals depend on plants: without the invisible harvest of all green plants, oxygen-breathing creatures would eventually perish. The oxygen requirements of two humans, for example, could be met by the output of a mature oak tree.

Another truly vital consequence of the gradual accumulation of oxygen molecules (two oxygen atoms linked together) as a constituent of the earth's atmosphere was the build-up of a layer of ozone molecules (three oxygen molecules linked together) in its upper reaches. Today this upper atmospheric ozone layer shields us from the full intensity of the sun's ultraviolet radiation. Until it was formed life was confined to the protection of oceans and lakes, and the dry land surfaces of the planet remained barren. With the evolution of oxygen-producing plants life on the land eventually became possible but it was a slow process. Beginning more than two-and-a-half billion years ago it was probably not nearing completion much before 500 million years ago. Meanwhile other innovations were being introduced.

One of these was sex. Before sex the multiplication of cells could only be achieved by the duplication of a single set of inherited DNA molecules followed by simple division of the rest of the cell

97

Life on earth

Life on earth is controlled by genes – coded instructions woven into the fabric of minute coils called chromosomes which make up the nuclei of living cells. As well as directing the activity of individual cells, genes decide the species an organism belongs to and most of its individual characteristics. They exercise their control through chemical reactions.

This diagram shows how the genes of a typical human cell make protein, one of the essential ingredients of our bodies. The genes for protein manufacture (1) are embodied in a sequence of chemical links between two interweaving strands of deoxyribonucleic acid, or DNA, which determines the order in which protein sub-units are assembled and therefore what kind of protein it is. The sequence is the code.

The process begins when the DNA strands unzip (2). The unpaired links then find fresh partners (3) from among the cell's reservoir of spare molecules to build a mirror image of the sequence – a new strand called messenger ribonucleic acid or messenger RNA. When that unzips (4), the pairing process is repeated, only this time the joins are made in threes (5),

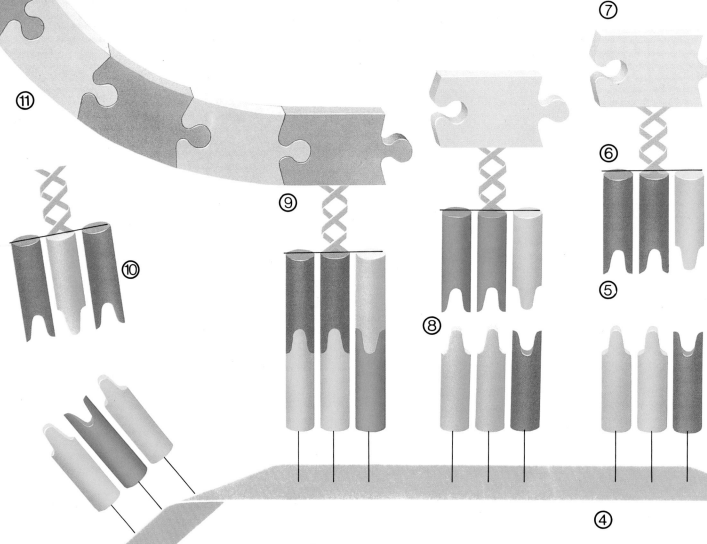

Separation of the chemical links which connect the two strands of a DNA molecule (2) allows the coded link sequence for protein production to be copied by a strand of messenger RNA (3). In this diagram the link, which might be one of two chemical pairs, is represented by a four-colour code – red linked with orange, and dark blue linked with light blue. The copy is a mirror image of the original.

A further 'reading' of the sequence, three links at a time (5), by a spiral unit of transfer RNA (6) determines the order in which different varieties of protein sub-unit (7) are added to the growing chain of the protein molecule (9).

98

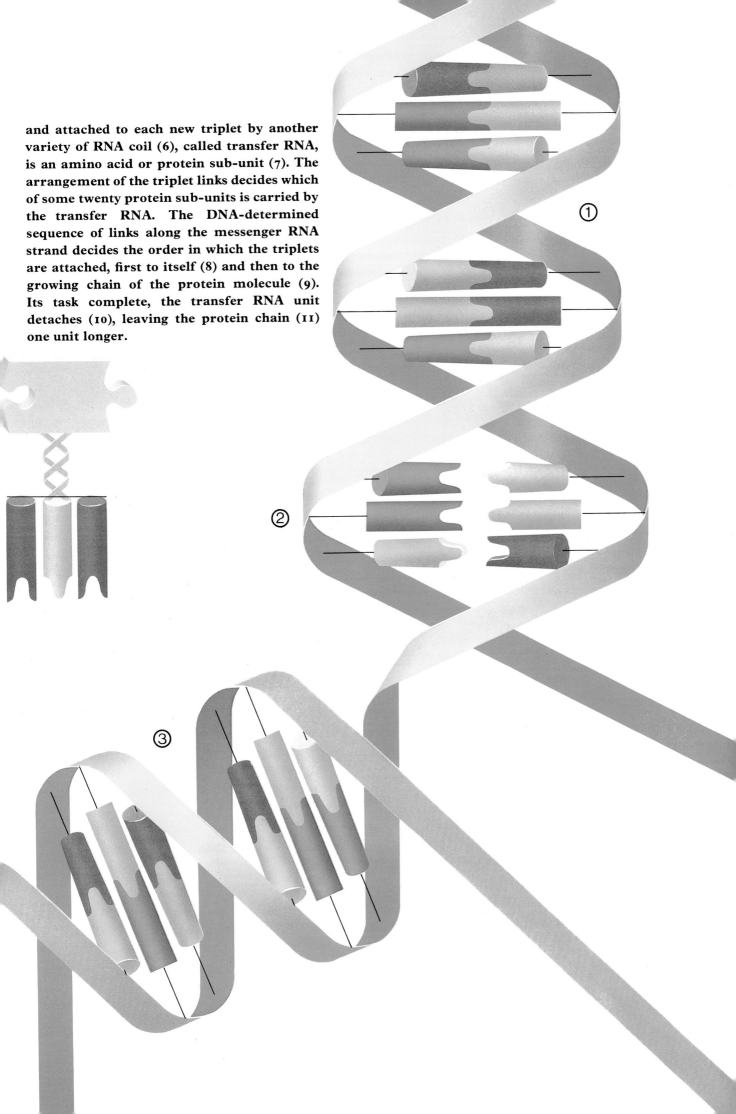

and attached to each new triplet by another variety of RNA coil (6), called transfer RNA, is an amino acid or protein sub-unit (7). The arrangement of the triplet links decides which of some twenty protein sub-units is carried by the transfer RNA. The DNA-determined sequence of links along the messenger RNA strand decides the order in which the triplets are attached, first to itself (8) and then to the growing chain of the protein molecule (9). Its task complete, the transfer RNA unit detaches (10), leaving the protein chain (11) one unit longer.

into two. And unless transcription 'errors' were made the resultant cell was identical to its parent. By involving two parents – each with a slightly different set of DNA molecules – a much greater variety of offspring became possible. For in sexual reproduction the genetic material of a new generation was now derived half from one parent and half from the other. Organisms of the same species which bred together produced a wider spectrum of varieties. The species was now less likely to be wiped out by natural selection – some varieties would almost always prove fitter than others. Another development was the extension of the rule of the DNA molecule beyond the boundaries of a single cell – the organization of cells into bodies. Just as the development of a molecular structure which could control others allowed the evolution of the cell, so the subsequent elaboration of a chemical messenger system, whereby the DNA molecules in some cells could be answerable to those in others, allowed the evolution of multicellular organisms.

The first of these cellular cooperatives were simple aquatic plants and animals structured like modern sponges, which obtain their nourishment by filtering the surrounding water. The composite cells gave each other physical support but little else. The next step saw the introduction of one of the fundamental elements of animal architecture – the cellular tube. In its simplest form it was probably closed at one end – a sack-like arrangement, two cells thick, with a single opening to let the surrounding water in and out. The inner layer caught the food particles and passed nutrients to the cells of the outer layer, which provided the community with a degree of protection.

The scavenging efficiency of these multicellular organisms was then greatly enhanced by the evolution of a second opening. With entry and exit points now separated, the tube became a literally streamlined food-processing production line. The animal body came to have a recognizable front and rear. The food-seeking end became the mouth. Encouraged by natural selection, cells were further specialized and organized into tissues with specific functions. For example, while networks of thread-like nerve cells provided coordination, powerful elastic cells making up muscular tissue improved an organism's mobility and thus extended the range of its foraging activities. With increasing sophistication of body design, internal cells – still essentially aquatic creatures but isolated from their watery ancestral environment by their neighbours – were provided with a system of irrigation canals – the precursor of the arteries and veins of the animal bloodstream.

Some 600 million years ago the configuration of the continental jigsaw puzzle was such that most of the dry land surfaces of the earth had come together to form a supercontinent in the planet's southern hemisphere. Unlike oceanic climates which are moderated by the capacity of water to store vast quantities of heat, continental climates tend to reflect the large seasonable variations in the amount of solar radiation they receive (that of Siberia, for example). Thus conditions in the shallow seas around the margins of this ancient barren land mass were probably very unsettled. The growth of unicellular aquatic plants became irregular and so therefore did the animal food supply. Among multicellular animal organisms the evolutionary response was more mobility and muscle power. Tissues were organized into limbs and, to survive lean times, some creatures took to rootling in the sea-floor mud for organic debris left over from periods of plenty. As animals became more bulky, natural selection promoted a number of structural innovations. The bodies of some – the ancestors of the insect world – became segmented while others – to protect and support their softer parts – grew shells or armour plates.

Some time around 550 million years ago the supercontinent split up. In the south Pacific, midway between the present locations of Easter Island and Tahiti, lay a land mass composed of much of the northern half of Europe. A continental segment composed of North America and Greenland filled the space now occupied by the northern half of that ocean. Wedged between South America, Asia, India, the Antarctic and Australia, Africa lay close to the South Pole. With the build-up of ozone in the upper atmosphere, the level of solar ultraviolet radiation reaching the surface was greatly reduced. The land became habitable.

Thanks to the moon, the first true land-dwellers were probably evolved from part-timers. For, along with the sun, the earth's seventy-four-quintillion-tonne satellite exerts

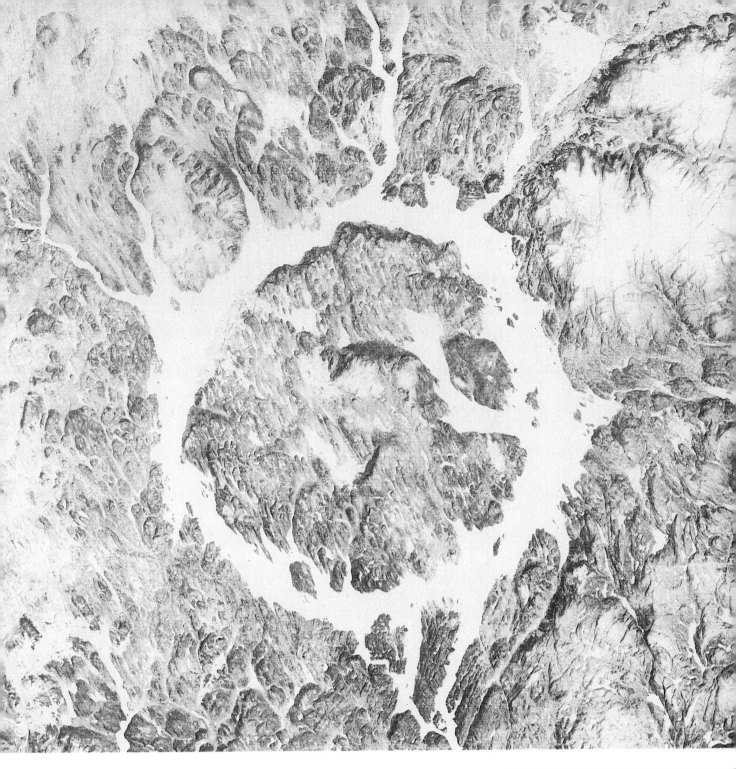

Like the rest of the inner planets, the earth was once pock-marked by craters formed during an early phase of solar system history when asteroids were more plentiful. Most of them were erased long ago by mountain-building and erosion. The one shown in the Landsat photograph above was preserved by a hundred metres of sediment and only revealed comparatively recently when this geological blanket was removed by ice-age glaciers. Today it forms the basin of Lake Manicouagan in Canada. Also from Canada, the microscopic single-celled fossil alga shown on the right is around 2,000 million years old.

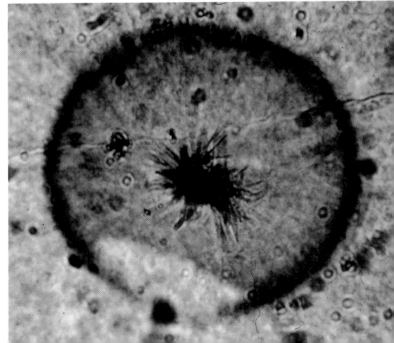

a considerable gravitational pull on its mother planet, distorting its liquid envelope so that there is always a slight bulge in the oceans on the face closest to the moon, where the attraction is strongest and a corresponding bulge on the opposite side, where the pull is weakest. The earth rotates but the bulges do not. Hence the ebb and flow of tides and the opportunity, some 450 million years ago, for natural selection to encourage organisms adapted to survive increasingly long periods of exposure to the atmosphere as they advanced ashore from the low-tide mark. With the beaches thus secured by bacteria, the invasion was followed through by plants and insects. In the oceans meanwhile, the protective function of bony armour-plating was giving way to its supporting role with the evolution of the internal skeleton.

By about 300 million years ago northern Europe had combined with Greenland and North America. The African-centred land mass had shifted substantially northwards as had the ancient nucleus of Asia. Associated with these rearrangements of the global patchwork, crustal plate collisions threw up fresh mountain ranges. The mountains attracted rain and rivers created extensive swamps which were soon covered with moss-carpeted forests of ferns and horsetails. Beneath the sinking forest floors peat beds, hundreds of metres thick, were formed by the continuous rain of plant debris from above. In time they would be overlain by river-borne sediments, compressed, and turned into coal. In the seas swam fish with eyes and gills, backbones and rudimentary brains. But on the mudflats and seashores between tidal rockpools, natural selection now favoured the replacement of fins by legs and gills by lungs to give rise first to the amphibians, which still return to the water to breed, and then to the reptiles, which do not.

The collision of continents continued. The African-centred land mass crossed the Equator to join the upper half of America and Europe which collided with Asia to produce the Ural Mountains. By about 250 million years ago it was possible to make a dry-land journey across this new supercontinent from California to Australia, or from the North Pole to the South Pole – though the last few hundred kilometres would be over an ice cap which had formed at the junction of the southern tips of South America, Africa, India

and where Australia met Antarctica. Elsewhere the climate was warm and dry. Desert and savanna conditions prevailed which favoured the reptiles over their amphibian rivals, who were more dependent on water. But with a return of wetter weather some 200 million years ago came a new green revolution. Plant food was once again plentiful and the herbivorous reptiles went on eating until some of their number weighed thirty tonnes.

So began the 130-million-year-long dynasty of the dinosaurs. At first these monsters were mostly vegetarians – *Brontosaurus* and its relatives for example. The carnivores came later with *Tyrannosaurus rex* and all that its name implies. But this king was to be the last of the line. Towards the end of his rule climatic changes caused natural selection to favour the evolution of warm blood, fur and, in place of the hazardous practice of leaving eggs around to be hatched by the warmth of the sun, it encouraged those creatures whose young were live-born and milk-fed. These small all-weather quadrupeds were the first mammals – an evolutionary line which would eventually lead to man. And although they may have been able to complete some of the steps – warm blood, for example – the dinosaurs died out. Their end was quite sudden. Their bones are found in rocks laid down at the close of what geologists call the Cretaceous period which ended about sixty-five million years ago. Bones are again found in the older formations of the next period – the Palaeocene. But they are not those of dinosaurs, they are the remains of mammals. Disease, a mammalian taste for dinosaur eggs, a swift global cooling of the climate, radiation from the explosion of a nearby star – all are possible explanations but exactly why the monsters died out while the mammals prevailed remains a mystery.

With the mammals came the first of the flowering plants. Among the greens and browns of forests and meadows, springtime brought brighter colours and autumn a fall of leaves. To attract their mates, insects mimicked the flowers and larger flying creatures exchanged drab leathery wings for flashy plumage. The continents too were on the move. As plate edges came together new mountain ranges were brought into being. The gradual collision of Africa and India with Europe and the rest of Asia gave rise to the Alps and the Himalayas – a process which continues today, as witnessed by frequent earthquakes occurring along the plate junctions stretching from Morocco in the west to Indonesia in the east. Meanwhile, the birth of the Atlantic brought the western edges of the Americas up against the Pacific Ocean plates, creating the coastal ranges of western North America and the Andes in South America. These too remain active, forming the eastern segment of the seismic and volcanic 'ring of fire' which encircles the Pacific.

Some twenty-five million years ago, among a great variety of mammals, the tree-dwelling common ancestors of apes and man were living in the forests of Africa, Asia and Europe. But then the opening of a new rift in the continental jigsaw puzzle, producing the Red Sea and flooding of the land along the present route of the Suez Canal, isolated the Asian members of this community. From this branch of the evolutionary tree, now separated from the main stem, came the arboreal orang-utans. In Europe and Africa our ancestors were beginning to spend less time in the trees. And on the ground they were spending less time on all fours. But in Europe and North Africa they faced competition in the shape of the bear – ancestor of the nursery variety but far from cuddly – which drove them south of the bear-proof Sahara.

Further rifting of the African land mass some twenty million years ago produced the valleys which still connect the Zambesi with the Nile. By fifteen million years ago these were flooded. To the west of this crocodile-infested barrier of lakes and rivers, the path of evolution led to gorillas in the forests and to chimpanzees in more open country. In the enclave to the east, between the crocodiles and the Indian Ocean, a drier climate saw the replacement of forests by grassland, meadows and savanna. Here – more than three and half million years ago – arose an upright two-legged creature with manipulative hands, endowed with a brain that gave him the transcendental powers of imaginative thought – *Homo*, perhaps not yet *sapiens*, but certainly man. Sociable and communicative, he emerged from his African home to become skilled in adapting, as well as adapting to, the varied environments of his planet, its moon and the space between them. And while today the instruments of man the toolmaker reach for other worlds, they also

Much of Canada lies under snow in this early springtime view of North America. It was taken by the Apollo 16 crew in 1972. The view may well have been much the same some 700,000 years ago, but at the end of a northern-hemisphere summer. For around then the climate cooled and an ice age, the first of the latest batch of eight, began. Although the Antarctic cap contains ten times as much ice as that of Greenland, the latter has the most pronounced effect on the world map during an ice age because it is surrounded by land while the former is in the middle of ocean. 18,000 years ago much of North America, northern Europe and Asia lay under several thousand metres of ice. Further south oak had given way to pine and tropical forests had become cold grassland. In the southern hemisphere, the lowered sea level may have created ice-age land bridges linking Australia with Asia.

Thinking up new theories to explain the 100,000-year cycle of ice ages is a popular scientific pastime. Perhaps the most attractive and one of the oldest, put forward in the 1920s and 1930s by the Yugoslavian geophysicist Milutin Milankovitch, relates the ice ages to long-term periodic changes in the tilt of the earth's axis and the shape of the orbit. There are three cycles at work: a 25,800-year swivel of the axis (so that the northern and southern hemispheres take turns to face the sun at closest approach), a 40,000 year 'nod' of the axis (the tilt varies between $21\frac{3}{4}°$ and $24\frac{1}{2}°$) and a 90,000-to-100,000-year variation in the form of the earth's orbit, between more and less circular (so that seasonal differences in the amount of sunlight falling on the planet wax and wane).

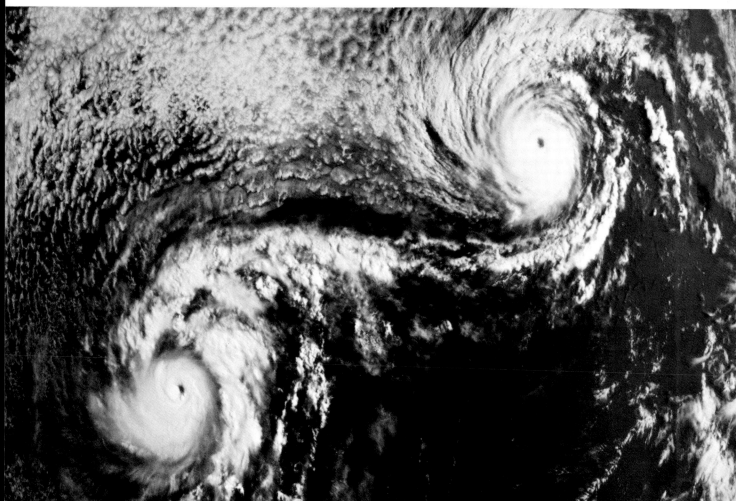

enable him to cast a global eye upon his own.

More than two thirds of its surface is ocean water. This reflects a blue atmosphere – four parts nitrogen (78 per cent) to one part oxygen (21 per cent). Were it familiar with the violent and continuous chemical reactions between oxygen and many other substances (which we call fire) an extraterrestrial intelligence searching for habitable planets would be greatly discouraged by this mixture. It also contains variable amounts of water vapour in the form of clouds. (If any living thing could escape destruction by fire, started by volcanoes or electrical storms, would it not then be crushed or drowned by tons of water falling from the sky?)

By comparison with that of Mars, for example, our atmosphere is dense and turbulent. Although planetary wind patterns are complex – modified in time and place by the alternation of day with night, the season and the nature of the surface (mountain, plain, desert, ocean) over which they blow – the movement of air masses is partly governed by a set of six circumplanetary bands similar to those of Jupiter and Saturn. The width of each of these bands is about 30° of latitude. They are arranged in three pairs. One member of each pair lies in the northern hemisphere, the other in the southern. Thus there are two trade wind bands, two bands of westerlies and two small polar bands.

Taken in August 1974 by an American National Oceanic and Atmospheric Administration (NOAA) satellite, the northern hemisphere cloud cover photomosaic above shows the true complexity of the earth's weather system. Drifting westwards within the northern trade-wind belt (see diagram overleaf), a string of hurricanes and tropical storms can be seen crossing the Pacific Ocean west of Mexico. These contracting, low-pressure air masses spin anti-clockwise in the northern hemisphere and clockwise in the southern hemisphere. (In the eastern hemisphere, hurricanes are called typhoons.) Expanding high-pressure air masses – such as those which descend from the poles – spin clockwise in the northern hemisphere and anti-clockwise in the southern hemisphere.

Another NOAA satellite image shows two of the hurricanes in detail. Both have distinct central eyes.

Within each band two main air-moving forces are at work. Warmed by the sun, convective turnover produces a surface flow of air towards the equator within the trade-wind bands. The same process produces an air flow towards the poles in the westerlies bands and away from them in the polar bands. At the same time the west–east rotation of the earth is having its effect. A mass of air moving towards the equator is also travelling away from the axis of the earth's rotation towards a region where the west–east linear motion is greatest. The west–east component of its speed will therefore be *less* than that of the surface which it is crossing. The result is an east–west wind, hence the trade winds. By the same rule – a result of the Coriolis force named for the French mathematician Gaspard Coriolis, who first thought of it while playing a game of billiards – the west–east speed of air masses moving polewards in the bands between 30° and 60° latitude is *greater* than that of the surface they are crossing – hence the westerlies. Between the bands are relatively calm zones – the doldrums around the equator and the horse latitudes at around 30° north and south. The latter are high-pressure zones formed by a pile-up of descending air, returning above the outflow, from the equator and latitude 60°. Deprived of its moisture content by rainfall at tropical and temperate latitudes, this is the dry air of the world's major deserts. High-pressure zones over the poles are the sources of expanding cold air masses which appear as 'highs' on the weather maps of temperate latitudes where Coriolis forces spin them clockwise in the northern hemisphere and anti-clockwise in the southern. In the case of contracting low-pressure air masses – hurricanes, most tornadoes and typhoons – the spin direction is reversed.

Above the weather, some 30 kilometres up, lies the earth's protective layer of ozone. Above this the atmosphere thins out. But situated at an altitude of about 1,000 kilometres there is another defensive shell – a magnetic doughnut which traps potentially harmful charged atomic particles, most of which come from the sun. First detected by satellites in the late 1950s, it is called the Van Allen belt after its discoverer, the American physicist James Van Allen. The belt is the most active part of the earth's magnetic envelope (see page 106). The equatorial doughnut arrange-

Seasons, moons and winds

The diagram on the right shows how the tilt of the earth's axis gives us seasons. At the spring equinox in March the northern hemisphere begins a six-month sunwards lean, which reaches its maximum with the summer solstice in June and ends with the autumn equinox in September. Southern hemisphere seasons mirror those of the north.

The phases of the moon depend on the relative positions of the sun, earth and moon. A new, crescent moon appears when the sun rises over the eastern limb of the near side.

The diagram of the earth's wind bands (below right) is a much simplified model of the planet's complex and fluid weather system. Between the bands of convection cells are zones of comparatively calm weather – the doldrums at the equator, for example. Here ascending trade-wind air (orange arrows) creates a zone of low pressure. The moisture it gathers is lost as rain as it spreads north and south over the tropics. Where it meets and descends with dry air from the westerlies bands (green arrows), it creates high-pressure zones – the calm seas and deserts of the horse latitudes.

The solar wind which blows across the earth's orbit (below) is partly deflected by the planet's magnetic envelope. The earth is also shielded from solar particles by a doughnut-shaped inner lining of the envelope called the Van Allen belt or belts. Above the poles, where there are holes in the doughnut, nocturnal lights are common when the sun is active.

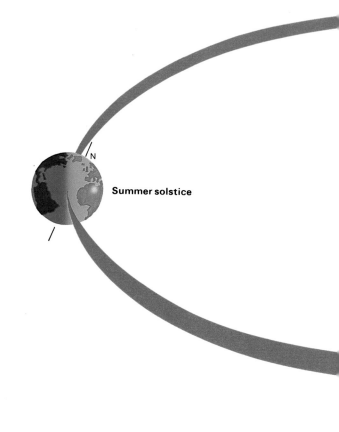

Summer solstice

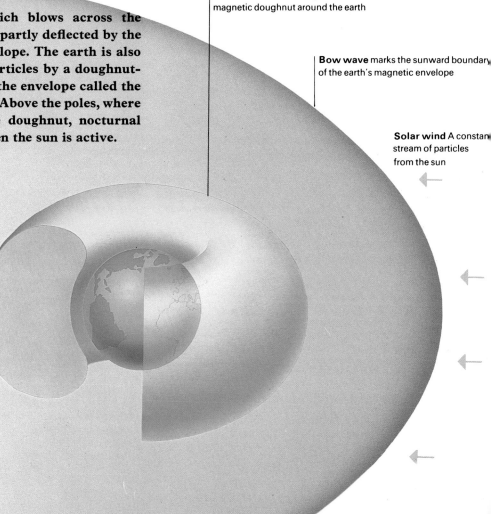

Van Allen belt forms a protective magnetic doughnut around the earth

Bow wave marks the sunward boundary of the earth's magnetic envelope

Solar wind A constant stream of particles from the sun

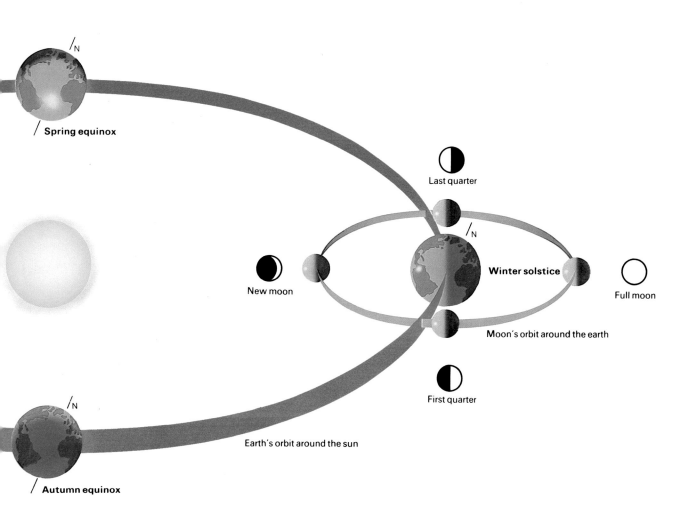

/N

Spring equinox

Last quarter

/N

New moon

Winter solstice

Full moon

Moon's orbit around the earth

First quarter

/N

Earth's orbit around the sun

/ **Autumn equinox**

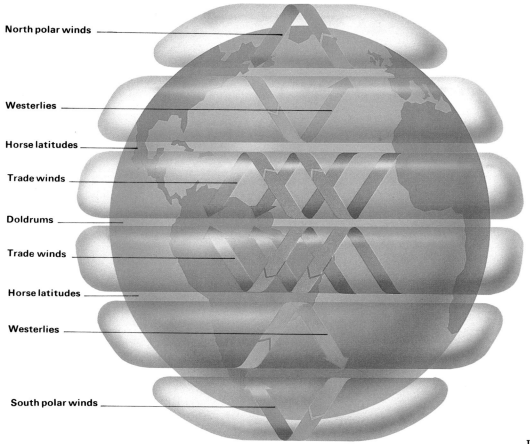

North polar winds

Westerlies

Horse latitudes

Trade winds

Doldrums

Trade winds

Horse latitudes

Westerlies

South polar winds

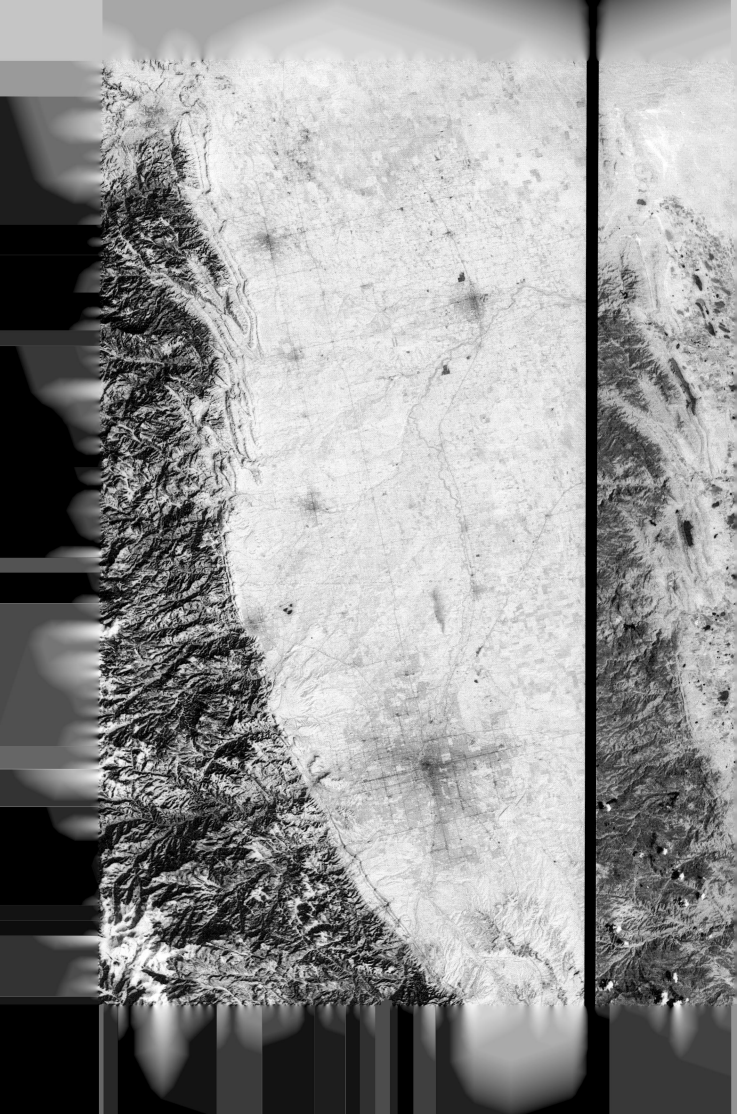

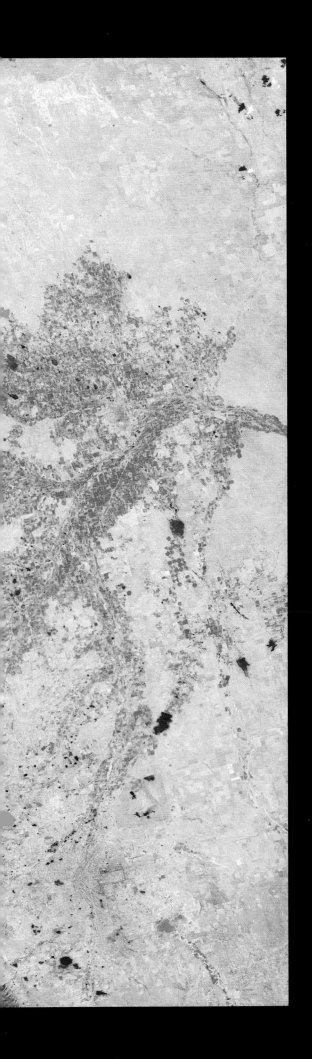

ment offers least protection to polar regions, and from above the Arctic and Antarctic charged particles from an active sun may be funnelled into the atmosphere. Here they may interact with atmospheric particles to give nocturnal auroral displays: Aurora Borealis in the northern hemisphere, Aurora Australis in the southern.

The earth's axis is tilted some $23\frac{1}{2}°$, which means that the amount of sunlight falling upon any part of the planet varies from day to day as it revolves around the sun – the northern hemisphere getting its maximum when leaning directly into the sun during the solstice in June, the southern hemisphere during the solstice in December. Thus the alternation of summer and winter between the hemispheres. When, at spring and autumnal equinox, the plane of the earth's equator passes through the sun, day has the same duration as night. Today the northern projection of the earth's axis points to Polaris, the pole star. But the planet's axis, while always tilted at about $23\frac{1}{2}°$, is itself slowing rotating. By around the year 15,000, the axial tilt will be such that the north star will be Vega and, were it not for the flexible nature of our calendar, northern hemisphere summers would occur in December and winters in June.

The earth's orbit is elliptical. At perihelion (its closest approach to the sun) in December it is five million kilometres closer than at aphelion (its greatest distance from the sun) in July. Because at perihelion the planet moves along its orbital track faster than it does at aphelion and because perihelion coincides with summer in the southern hemisphere, this season should be hotter and shorter than its northern counterpart. And in theory it *is* four days shorter. But by absorbing

---

North and south of the tropics, the face of the earth alters according to the season. Winter (left) and summer (right) along the eastern edge of the Rocky Mountains in the United States are seen here in two Landsat images taken in January and July 1973. The large dark patch in the snow (lower centre, left-hand image) is the city of Denver.

In this kind of satellite photograph infra-red radiation from plants, as well as visible light, is recorded. Thus pink and red patches in the summer image denote farmlands (on the plains), pine trees and alpine meadows (in the mountains).

heat at the start of the southern summer, the ocean water making up four fifths of the surface of that hemisphere limits the rise in temperature and, by losing heat gradually at the end of the summer, effectively extends that season so that, in practice, the difference is hardly noticeable.

Four fifths of the southern hemisphere and three fifths of the northern hemisphere are under water. The volume of the world's oceans is eleven times greater than that of land above sea level. And thanks to the thermal storage capacity of water, the oceans which make up the hydrosphere are rarely cooler than 0 °C or warmer than 30 °C. Thus the whole of the hydrosphere is also part of the habitable part of the planet, or biosphere. But of the dry-land volume, only some of its surface (with shade temperatures ranging from *minus* 53 °C to 58 °C), and the part a little way below and a little way above it belong to this category. By volume therefore, the biosphere is almost entirely under water – a fact reflected in the distribution of life on this planet. Of some 450,000 species (excluding insects) which constitute the animal kingdom, for example, four fifths make their homes in the seas.

Insulated by sheer volume from the more extreme effects of seasonal variation, the hydrosphere is nevertheless prone to long-term temperature change. Within the last 700,000 years eight cycles of depressed (by as much as 15 °C) global average temperature have resulted in a vast increase in the extent and thickness of the north polar ice cap at the expense of sea level and water temperature. How these and earlier ice ages came about (and whether the planet is still recovering from a past, or already entering a new, period of glaciation) remains, like the

Built-up vegetation-free areas appear blue-grey in Landsat images. This one, taken in October 1972, shows Los Angeles (peninsula, bottom right). The sharp linear boundary, running from left to right across this scene, between the mountains to the north of the city and the Mojave Desert (large light area, upper right), is the San Andreas Fault which lies along one of the earth's major crustal plate junctions (see diagram, pages 92–3). The bleached areas in the middle of the desert are dried-up lake basins. One of them has been used as a runway for Space Shuttle landing tests.

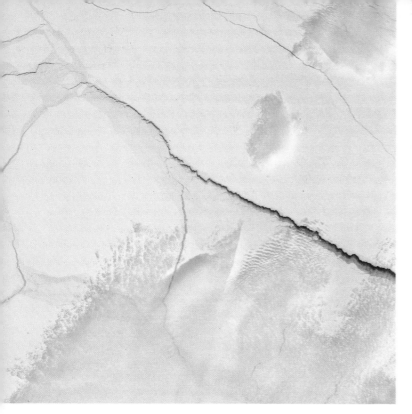

More than two thirds of the earth's surface is under water. Some of it, like the 34,000 square kilometres of the Beaufort Sea shown above, is frozen. About one fifth of the land surface, like the 34,000 square kilometres of sand dunes in Arabia's Empty Quarter shown below, is desert.

Fresh water and desert are united in the Landsat image (right) of the Nile Valley and Delta. One of mankind's oldest settlements, dating back some ten to twenty thousand years, to when a drying climate encouraged nomadic hunters to become farmers, the irrigated Delta (red) is today the home of about 30,000,000 Egyptians. Cairo lies at the apex of the Delta where the Rosetta Nile (left) leaves the Damietta Nile (right).

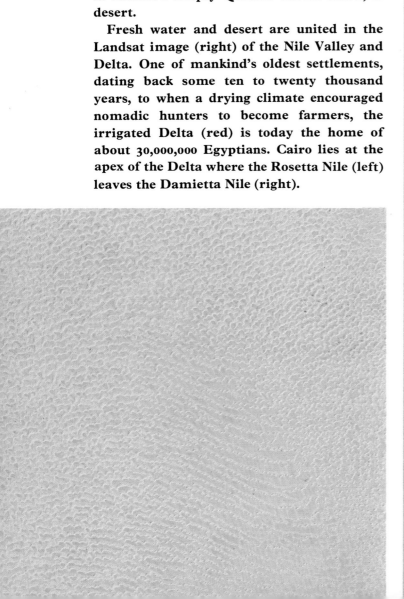

A relatively cloud-free view (above), taken by a United States National Oceanic and Atmospheric Administration weather satellite in 1976, shows Britain, Ireland and the coast of the European mainland from Brittany (lower edge, centre) to Denmark (top right).

The Landsat image on the right, taken in July 1975, shows London (large blue patch) straddling the River Thames (lower centre and right). West of the city the bleached area near a number of reservoirs (dark blue) is Heathrow Airport. Just south of the Thames, an irregular L-shaped mark which looks like a scratch, or the result of a small hair on a photographic negative, is a recent cutting in chalk hills to accommodate new sections of motorway.

The bent-sausage shape of the Serpentine (dark blue) – the boating lake in Hyde Park (pink) – can be seen in the centre of the detail showing the London area (below).

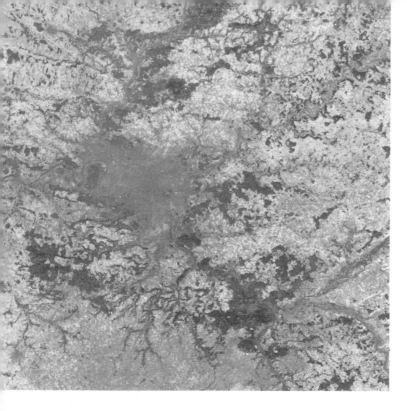

Paris (above) in late winter, Copenhagen (centre, right) in late summer and Munich (below) in early autumn.

Woven into the meanders of the River Seine, Paris (light blue) is surrounded by forests (dark red) and open farmland (pale and pink at this time of the year). One of the most intensely farmed areas on earth, the Danish island of Zealand occupies the centre of the image on the right. Copenhagen lies on its eastern coast facing Sweden across the Öresund. The Swedish lowland of Skåne can also be seen.

Munich is surrounded by forests (dark red) and farmland which clearly flourishes well (bright red) in fertile river valleys.

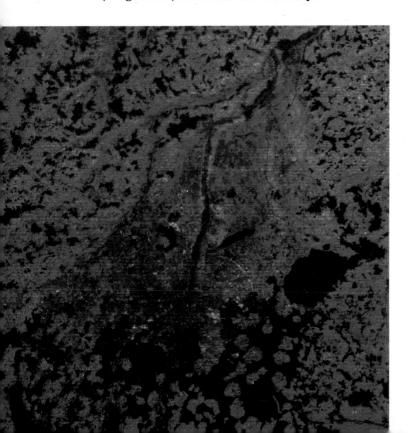

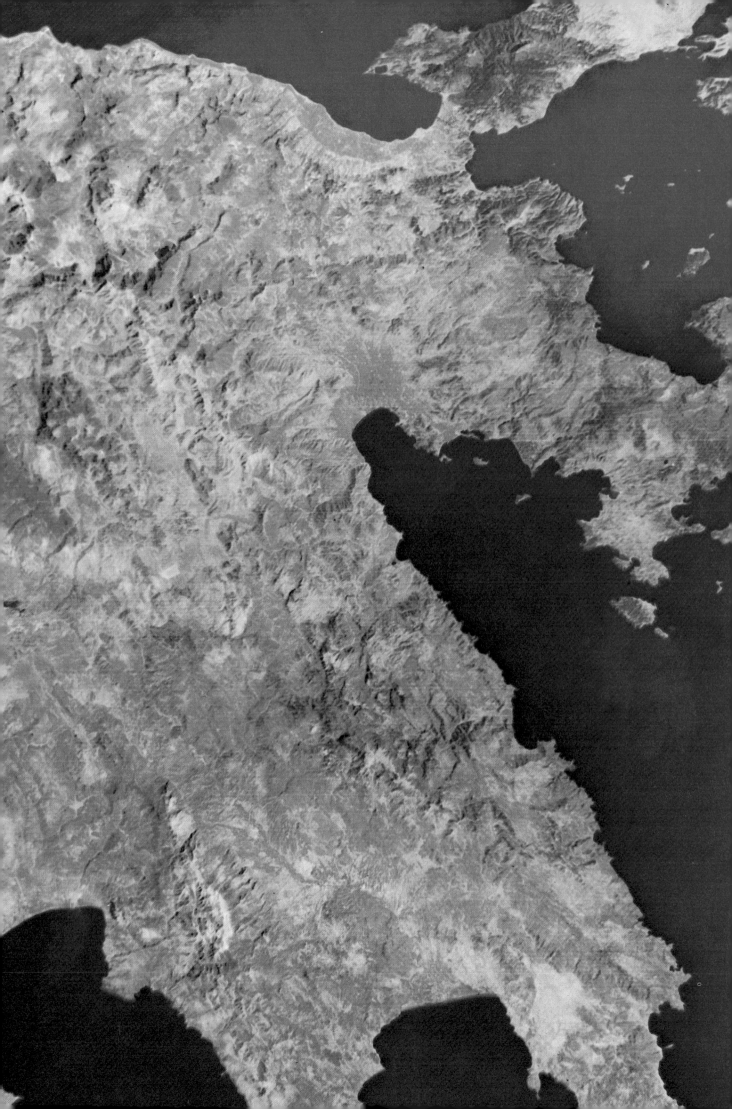

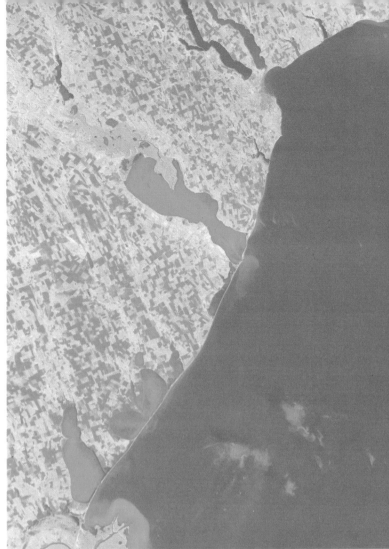

The bold patchwork of the huge cornfields along the Black Sea coast of the Ukraine (above) contrasts strongly with the rough and ancient pastures of the Peloponnese (left).

The large lagoon-like bay in the centre of the Landsat image above is the mouth of the River Dnestr. The port of Odessa lies in the C-shaped bay in the bend of the coast at the top.

The image of southern Greece, taken in August 1972, includes Athens (blue patch, top right). To the west of the capital is the isthmus which joins the Peloponnesian peninsula with the rest of Greece. This is cut by the Corinth Canal (thin blue line) which joins the Gulf of Corinth (top left) with the north-western corner of the Aegean Sea (right). A mountainous area, it is part of the Alpine chain raised by the collision of Europe with Africa (see diagrams, pages 92–5). Originally deforested by man, it is kept that way by goats and sheep. Low rainfall and porous limestone make this a difficult region for farmers. A number of rich pink patches at the head of some bays denote irrigated areas where fruits and vegetables are cultivated.

demise of the dinosaurs, a controversial question. But cycles of solar activity and changes in the earth's orbit will probably form part of the answer.

Although only the sixth largest in the solar system, the earth's moon is by far the largest satellite in proportion to the dimensions of its mother planet – about a quarter of its diameter and about one eighteenth of its mass. And their relationship is close – lunar rhythms are both dynamic and biological. It was once closer. Today the moon is spiralling away from the earth at a rate of about three centimetres a year. Its present average distance is 384,400 kilometres. At the same time the lunar pull upon the world's oceans is having a frictional effect on the earth's rotation, slowing it down and lengthening the day at a present rate of about sixteen seconds every million years. It has been suggested that when the earth–moon distance reaches around 480,000 kilometres, the length of an earth day (currently 23 hours, 56 minutes, 4 seconds sidereal – from fixed star to fixed star) will then become the same as a lunar month (currently 27·3 earth days sidereal). As well as the ocean tide, the moon also produces a land tide. And it is possible that this regular deformation of the earth's geometry by as much as 20 centimetres plays a part in triggering seismic activity along the margins of crustal plates. If so the better understanding of such a relationship would play an important part in the prediction of the location and timing of earthquakes.

Calculations involving shifts in the lunar orbit and changes in the earth's period of rotation have been extended backwards as well as forwards in time. In the late nineteenth century the English mathematician George Darwin (son of Charles) computed that when the earth–moon distance was zero, the earth day would have been about four hours long. His suggestion that the moon was once a part of the earth is but one of the three principal contending theories concerning the origin of our lunar companion.

The current version of Darwin's hypothesis is that when the earth became molten some 4·6 billion years ago, settling of the heavier metallic content to form the core around the centre of mass increased the planet's rate of rotation (the pirouetting ballerina effect) until the day was only 2 hours 36 minutes long. At this point centrifugal forces, acting upon this spinning globe of fluid rock, produced an irregular equatorial bulge in the mantle blanket. And – as anyone with experience of a potter's wheel will know – the whole then became extremely unstable. The earth underwent a series of rapid morphological transformations – from a squashed sphere to an egg spinning on its side and from an egg to a pear. The moon was born when the top of the pear spun off into earth orbit.

Supporters of this 'daughter' hypothesis cite the average density of lunar material (3·3) which is about the same as that of uncompressed rocks forming the outer layers of the earth whence they suppose it was drawn. The chemical composition of moon rock is, however, markedly different – samples brought back by Apollo astronauts are noticeably richer in aluminium, calcium and titanium but poorer in the more volatile elements like silver, zinc and gold than are comparable earth rocks.

The second theory of lunar origin has the moon formed separately but at the same time and place as the earth, as a sister planet. This idea too has been somewhat shaken by the results of moon rock analyses. To explain how the earth and moon came to have different chemical compositions – although derived from the same dust and gas mixture in one of the rings surrounding the primitive sun – patrons of this 'sister' hypothesis have suggested that the earth was the first to form and that the moon was then put together by accretion of chemically altered (by the tremendous heat radiated from the surface of the young and molten earth) low-density leftovers.

Theory number three supposes that the moon was born elsewhere and that its present situation is the result of gravitational attraction and capture by the earth – the 'wife' hypothesis. Its adherents point out that the solar system is full of

Full moon over the Libyan Sea from the south coast of Crete. Of seven solar system moons with diameters greater than 3,000 kilometres, the earth's has the smallest 'parent'. Their relationship is reflected in the twice daily ebb and flow of tides and in longer-term biological cycles linked to the lunar month. Some 4·6 billion years old, it was lifeless until July 1969 when Apollo 11 landed in the Sea of Tranquility, which lies to the right of the centre of the near-side face.

moon-like objects. Our moon, Europa and Io (two of Jupiter's satellites) for example are almost triplets in terms of mass and density. But the chemistries of these Jovian moons, other astronomers argue, may prove very different to that of our own, the constituents of which, they believe, came from a region of space much closer to the earth. Another version of this theory has the moon assembled in earth orbit by accretion of a large number of separately captured moonlets.

The subsequent history of our lunar companion has been fitted together from evidence brought home between July 1969 and December 1972 by the crews of the six successful Apollo missions, from data returned by the instruments they left behind on the moon's surface and from photographs and other information gleaned by unmanned American Lunar Orbiter, Surveyor and Russian Luna spacecraft. While the exact manner of the moon's arrival in earth orbit is in dispute, the date of its birth is not. The moon was born some 4.6 billion years ago along with the earth and the rest of the planets and their moons. Shortly afterwards the outermost few hundred kilometres of our 3,473-kilometre-diameter satellite became molten – probably (assuming that the moon was formed by accretion) as a result of the conversion of impact energy into heat. As its surface cooled again a crust was formed.

In places subsequent re-melting of part of the crust produced the molten form of a rock called KREEP, a uniquely lunar rock. Compared to the rest of the lunar crustal rock, KREEP is rich in potassium (chemical symbol K), the rare earth elements (initials REE) and phosphorus (chemical symbol P). KREEP is also comparatively rich in the radioactive elements thorium and uranium. Detectors carried by the 1971 Apollo 15 and 1972 Apollo 16 spacecraft mapped several KREEP occurrences in the vicinity of the crater Copernicus.

The formation of the lunar crust was accompanied by a steady hail of meteoroids which pitted the young moon with craters. Parts of this early rugged moonscape are still to be found today, notably on the face of the moon turned away from the earth. Then, between around 4.2 and 3.3 billion years ago, lunar gravity attracted a number of much larger objects. The basin of the Sea of Tranquility or *Mare Tranquillitatis*, for ex-

ample – site of the first manned landing (Apollo 11, July 1969) – is the scar left by one of the earliest of these dramatic encounters.

The largest such near-side crater – 1,050 kilometres from rocky rim to rim – is now the basin of the Sea of Rains or *Mare Imbrium*. One of the eyes of the 'Man in the Moon', it was formed some 3.8 billion years ago by a colliding asteroid with an estimated diameter of seventy kilometres or more. The impact excavated several thousand cubic kilometres of crust forming a vast fireball of glowing gas, vapour and shattered rock fragments which fell back to the lunar surface, coating it with an 'ejecta' blanket of rubble and numerous secondary craters. An even larger basin – some 2,000 kilometres in diameter – was revealed by altimeters aboard the Apollo 15 and 16 spacecraft in the southern sector of the hidden side of the moon. Also often partly hidden is the Eastern Sea or *Mare Orientale* basin. About 900 kilometres across and one of the youngest of these large-scale moonscape features, it was formed around 3.8 billion years ago.

While the face of the moon was being cratered and crumbled its interior was being slowly heated by the decay of its radioactive components. By around four billion years ago large pockets of molten rock had formed beneath the crust. Where this was deeply cracked – where it underlay the huge impact basins – the lava found an easy route from the mantle to the surface, flooding the larger craters to give the moon its relatively smooth *maria* or 'seas'. But although large craters (with diameters of 200 kilometres or more) are fairly evenly distributed around the lunar globe, nearly all the moon's 'seas' are located on the earth-facing side. Only two far-side basins – those of the Sea of Moscow or *Mare Moscoviense* and the crater Tsiolkovsky – are filled with lava. A thicker far-side crust, it is suggested, prevented the lava from reaching the surface.

By the time the last of these great flows had poured across the lunar surface – perhaps around a billion years ago – the earth too had acquired its oceans, its rhythms of rain and erosion, its cycles of sedimentation and mountain building. But on the moon there was no rain, the red glow of lava faded, the seas set solid and so did the mantle beneath them. Unlike that of the earth, the lunar crust was now immobilized, the gaze of the Man in the Moon forever fixed.

Craters large and small. The view from a Lunar Orbiter spacecraft (below) shows the lava-filled basin of the Moscow Sea, or Mare Moscoviense (left). About 350 kilometres across, it lies in the northern part of the far-side hemisphere. Another Lunar Orbiter image (right) shows Plato. Some 100 kilometres across, it is also lava-filled and lies to the north of the Sea of Rains, or Mare Imbrium. The colour photograph taken by an Apollo 10 astronaut in May 1969 shows Keeler, a 150-kilometre-diameter far-side crater with central peaks. The small crater (above), photographed by an Apollo 16 astronaut, is 900 metres across.

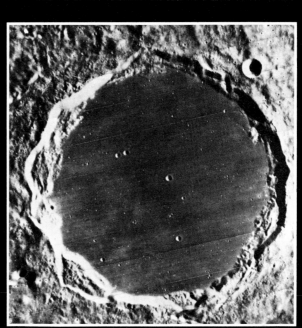

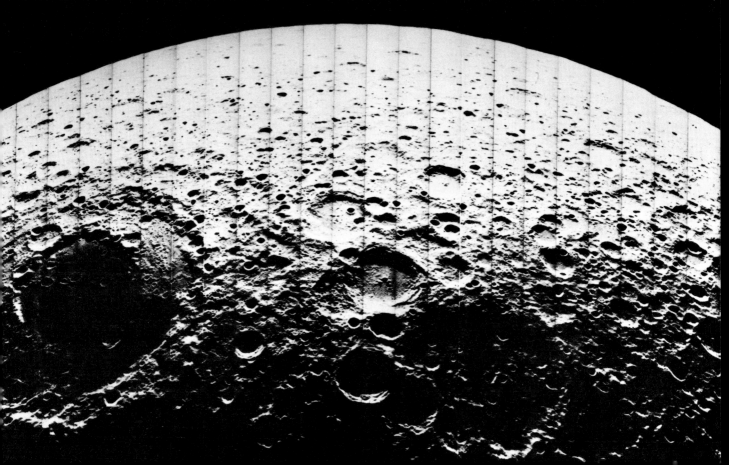

The stopped clock face of geological time, his stare – although splattered in places by a few 'young' craters of some size and generally freckled by many of much smaller measure – has been much the same for millions of years.

One of the youngest of these late arrivals is the crater Tycho. Imprinted among the much older basins of the near-side southern highlands some 100 million years ago, it is surrounded by rays of ejected material extending for hundreds of kilometres in all directions. Its formation must have been a spectacular event. Another conspicuous 'recent' rayed crater is that of Copernicus. To the south of the southern rim of Mare Imbrium, it was cut into the lunar surface some 900 million years ago when the earth, above sea level, was as lifeless as the moon. To the north-east of Copernicus, in the mountainous margin of Mare Imbrium, is Eratosthenes. Older than either Tycho or Copernicus but younger than the surrounding 'seas', the rays of this crater have been largely worn away by the relentless churning of the lunar surface by billions of small meteoroids.

Those regions of the moon's surface not covered by lava 'seas' – much of the southern part of the near-side face and most of the far side – are often collectively called the 'lunar highlands'. Among them are a number of impressive peaks. Like those of the earth they are the product of colliding rock masses but unlike those of the earth, the construction of which took millions of years, they were built in a matter of minutes – for the mountains of the moon are mostly the walls of impact craters. A typical example is the range making up the south-eastern rim of Mare Imbrium (and the lower part of the nose of the Man in the Moon), the Apennines, or *Montes Apennini* (their geographical form resembles the outline of the Italian peninsula); they stretch more than 950 kilometres from the crater Eratosthenes in the south to Mount Hadley which rises nearly 1,900 metres above the surrounding maria in the east.

To the west of Mount Hadley, among the Apennine foothills which mark the eastern end of the part of Mare Imbrium called the Marsh of Decay or *Palus Putredinis*, is a sinuous channel – the 135-kilometre-long Hadley Rille or *Rima Hadley*. On average 1,200 metres from bank to bank and 370 metres deep, it looks from above a little like a river bed, which is probably what

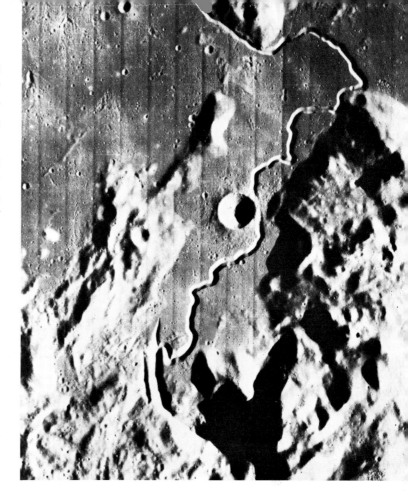

The famous 1966 Lunar Orbiter photograph on the left gives an oblique view of the walls and central elevations of the crater Copernicus as seen from the south. Insets above it show the location (boxed, left), rim and central peak (detail, right) of the rayed crater Tycho.

Another Lunar Orbiter image (above) shows the 135-kilometre-long Hadley Rille or Rima Hadley. The bottom of this 370-metre-deep channel can be seen in a photograph (below) taken by an Apollo 15 astronaut in 1971.

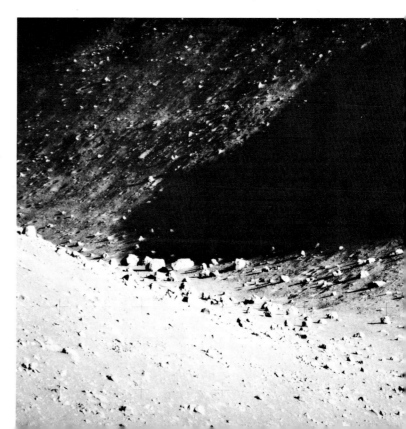

Moon, near side

A composite of several images taken at different phases of the moon, this photograph shows the main near-side maria and craters.

Outlined in the geological interpretation on the right are impact basins with diameters greater than 220 kilometres.

In the following list of formations shown on the geological map, the location of each formation is indicated by a key number followed by an indication of the quadrant in which it is found: north-west (NW), north-east (NE), south-east (SE) or south-west (SW).

Craters

Agrippa	34 (NE)	Manilius	30 (NE)
Alphonsus	51 (SW)	Maskelyne	32 (NE)
Archimedes	46 (NW)	Plato	33 (NW)
Aristarchus	5 (NW)	Plinius	29 (NE)
Aristillus	7 (NE)	Stevinius	39 (SE)
Aristoteles	24 (NE)	Taruntius	36 (NE)
Atlas	48 (NE)	Theophilus	35 (SE)
Bullialdus	17 (SW)	Timocharis	6 (NW)
Cassini	49 (NE)	Tycho	18 (SW)
Cavalerius	19 (NW)		
Copernicus	9 (NW)		
Eratosthenes	8 (NW)		
Eudoxus	25 (NE)		
Geminus	27 (NE)		
Hercules	26 (NE)		
Kepler	10 (NW)		
Langrenus	37 (SE)		

Lacus ('lake')

Lacus Somniorum (Lake of Dreamers)	23 (NE)

Maria ('seas')

Mare Australe (Southern Sea)	44 (SE)
Mare Crisium (Sea of Crises)	28 (NE)
Mare Fecunditatis (Sea of Fertility)	40 (SE)
Mare Frigoris (Sea of Cold)	16 (NE)
Mare Humorum (Sea of Humours)	3 (SW)
Mare Imbrium (Sea of Rains)	1 (NW)
Mare Marginis (Marginal Sea)	43 (NE)
Mare Nectaris (Sea of Nectar)	38 (SE)
Mare Nubium (Sea of Clouds)	2 (SW)
Mare Orientale (Eastern Sea)	45 (SW)

Mare Serenitatis (Sea of Serenity)	47 (NE)
Mare Smythii (Smyth's Sea)	42 (NE)
Mare Spumans (Foaming Sea)	41 (NE)
Mare Tranquillitatis (Sea of Tranquility)	31 (NE)
Mare Vaporum (Sea of Vapours)	22 (NE)

Montes ('mountains')

Montes Apennini (Apennine Mountains)	11 (NE)
Mount Hadley	12 (NE)

Oceanus ('ocean')

Oceanus Procellarum (Ocean of Storms)	4 (NE)

Palus ('marsh')

Palus Putredinis (Marsh of Decay)	13 (NW)

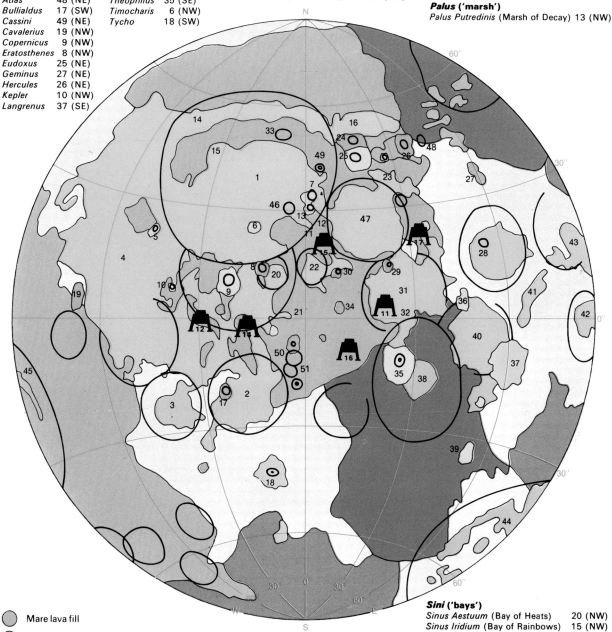

- Mare lava fill
- Imbrian basins
- Nectarian basins
- Cratered
- Heavily cratered
- Eratosthenian craters
- Copernican craters

Apollo mission number and landing site

Sini ('bays')

Sinus Aestuum (Bay of Heats)	20 (NW)
Sinus Iridium (Bay of Rainbows)	15 (NW)
Sinus Medii (Central Bay)	21 (NW)
Sinus Roris (Bay of Dews)	14 (NW)

The largest near-side lunar feature, the right eye of the Man in the Moon, is Mare Imbrium (1). Now a calm 'sea' of dark lava (see photograph), its basin was excavated by a colliding asteroid some 3·8 billion years ago. The left eye, the 4·1-billion-year-old basin of Mare Serenitatis (47), is also lava-filled.

In the map above the lunar surface has been classified (see key, left) into generally ancient (dating from more than 4·2 billion years ago) cratered and heavily cratered zones, Nectarian basins (created by collisions contemporaneous with that which formed the basin of Mare Nectaris (38) between 4·2 and 3·8 billion years ago), Imbrian basins (created by impacts between 3·8 and 3·3 billion years ago) and lava-filled mare regions (with a considerable range of ages from ancient to modern). A number of large, relatively young Erastosthenian (between 3·3 and 1·8 billion years ago) and Copernican (from up to 1·8 billion years old) craters are also shown.

Moon, far side

The Apollo 16 photograph on the right shows part of the moon's far side where the hidden half of the eastern hemisphere (right) meets the near side (left). The geological interpretation below (see also previous page) reflects various phases of cratering.

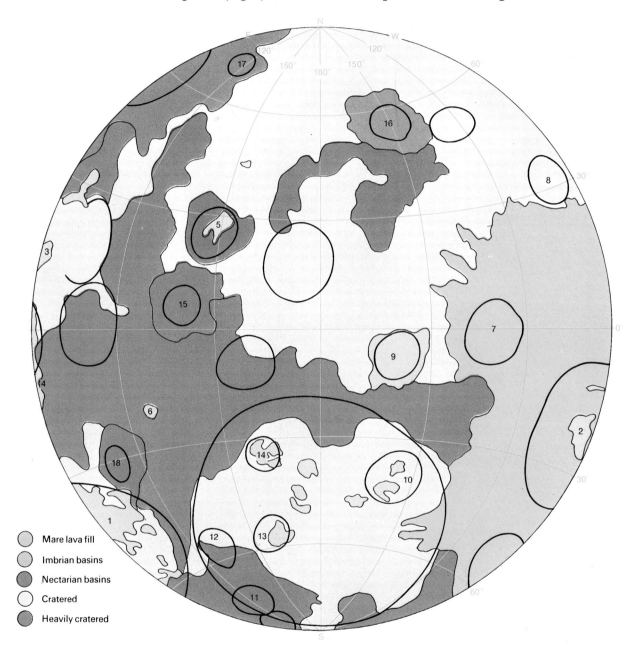

Mare lava fill

Imbrian basins

Nectarian basins

Cratered

Heavily cratered

In marked contrast to the near-side face, the hidden hemisphere of the moon is far less affected by lava flows. Perhaps because it has a thicker crust, features on the far side are almost exclusively the work of impacting bodies.

To the east of Mare Crisium (left edge of photograph), the dark lava-filled basins of Mare Marginis (3) and Mare Smythii (4) can be seen. The crater Tsiolkovsky (6) lies close to the terminator (lower right edge of the photograph).

In the following list of formations shown on the geological map, the location of each formation is indicated by a key number followed by an indication of the quadrant in which it is found: north-east (NE), north-west (NW), south-west (SW) or south-east (SE). Note that on this map the eastern hemisphere appears on the left and the western hemisphere appears on the right.

Craters

Apollo	10 (SW)	Poincaré	13 (SE)
Birkhoff	16 (NW)	Schrödinger	11 (SE)
Hertzsprung	7 (SW)	Schwartz-	
Korolev	9 (SW)	schild	17 (NE)
Lorentz	8 (NW)	Tsiolkovsky	6 (SE)
Mendeleev	15 (NE)		
Milne	18 (SE)		
Planck	12 (SE)		

Mare ('seas')

Mare Australe (Southern Sea)	1 (SE)
Mare Ingenii (Sea of Engineers)	14 (SE)
Mare Marginis (Marginal Sea)	3 (NE)
Mare Moscoviense (Moscow Sea)	5 (NE)
Mare Orientale (Eastern Sea)	2 (SW)
Mare Smythii (Smyth's Sea)	4 (SE)

A huge lunar boulder (above) dwarfs an Apollo 17 astronaut. An Apollo 16 astronaut (below) stands beside a lunar roving vehicle. The Apollo programme included six moon landings – two in 1969 (Apollos 11 and 12), two in 1971 (Apollos 14 and 15) and two in 1972 (Apollos 16 and 17). Lunar roving vehicles were used on the last three missions. Between them the Apollo landings yielded nearly 400 kilograms of lunar samples, thousands of photographs and much other on-site data. Samples have also been returned by Russian Luna craft – Lunas 16 (in 1970) and 20 (in 1972), which landed in the Mare Fecunditatis region, and Luna 24 (in 1976), which landed in Mare Crisium. The moon's surface has also been explored by remotely controlled Lunokhod vehicles – Lunokhod 1 which was deployed by Luna 17 in 1970 to cover $10\frac{1}{2}$ kilometres in Mare Imbrium and Lunokhod 2 which was delivered by Luna 21 in 1973 to travel 37 kilometres in Mare Serenitatis.

it is. But the liquid it once carried was lava, not water. Rima Hadley is but one of a large number of such channels cut into lunar flow surfaces. Elsewhere, around the margins of the larger 'seas', extensive sinuous ridges – often in the form of concentric rings – run parallel to and just inside the basin rims. Recalling the wrinkled skin of dried fruit, they are thought to have been formed in much the same manner – by crustal shortening as the volume of lava contracted beneath them. In other places rough- or smooth-flanked dome-shaped hills, up to 20 kilometres in diameter, rise several hundred metres above the surrounding 'sea' plains. Like their terrestrial equivalents they are thought to have been pushed up from below by ascending lava bodies.

Ten degrees above the boiling point of water at lunar noon, the surface temperature falls to more than 140 °C below zero just before dawn. But when the sun does come up, its steely light falls on ice-free slopes, for there is no water on the moon. And all day and every day the same litany of shadows plays across its ancient surface. Only very rarely does the flicker of expression cross its solar-wind-polished meteoroid-scoured face. The observation of transient 'flashes', 'glows' and 'mists' dates from the eighteenth century. Their site, more often than not, has been the crater Aristarchus in the northern part of the Ocean of Storms, or *Oceanus Procellarum*, near the north-western edge of the near-side face. While successive Apollo missions found no evidence of current volcanic activity, they did detect – from orbit – a number of radioactive regions, including places where the rare gas radon was being vented from the surface (one of these was Aristarchus) but this is thought to be a purely local phenomenon. Elsewhere the crust and the mantle beneath is considered to be geologically 'dead'. Deeper still the story may be different. One of the instruments taken to the moon by the Apollo astronauts was a device for measuring the temperature below the surface. This, it was found, does not vary like that of the exposed surface but remains constant at any given level with an increase in depth. The centre of the moon, it was suggested, must be quite hot. Then in July 1972 the near-side Apollo seismographs detected the impact of a one-tonne meteoroid. It had crashed on the far side – so the shock waves had reached the recorders after a journey through the centre. From their behaviour it was deduced that the moon has a hot, partially molten, possibly metallic, core over 1,000 kilometres in diameter.

Our lunar companion takes 27·3 days to complete one circuit of the earth with reference to the stars – a length of time called the sidereal month. This is exactly the same in both direction and period as its axial rotation, so only one hemisphere of the moon ever faces the earth. Like the motions of Mercury, this is an example of what astronomers call spin-orbit coupling. Because the earth-moon system is moving around the sun, the interval between new moons – a length of time called the lunar month (which is exactly the same as the interval between sunrises on the moon – the lunar day) – is slightly longer at 29·5 days.

Although there are biological cycles related to both, there is no simple relationship between the lengths of the lunar month and the earth year, which is 12·4 times longer. (There is however a calendar in which the 'year' consists of exactly twelve lunar months – that of Islam. The Muslim 'year' is eleven days shorter than the earth's orbital period. Thus their months slip backwards through the seasons. In 1977, *Hijri year 1397*, the month Ramadan began in late (northern hemisphere) summer. In 1987, *Hijri year 1407*, it will start in the spring).

The present earth–moon distance varies between 356,410 and 406,740 kilometres. Once a month the moon passes between the earth and the sun, but because the plane of the moon's orbit around the earth is tilted at an angle of 5° to that of the earth's orbit around the sun, the three bodies are only occasionally in perfect alignment. Only when the moon is at a point where these two planes intersect do we have a *solar eclipse*. *Lunar eclipses* occur when the shadow of the earth falls on the moon.

In July 1969 the first two of twelve Apollo astronauts landed on the moon. The analysis of the sizeable pile of rocks and soil they brought back has involved several hundred scientists from all over the world. Their work is far from complete and a large part of the Apollo moon material will be kept back so that a new generation of investigators with fresh techniques will have untouched samples to draw upon without the enormous expense of going to the moon to get them.

A thick blanket of dust and debris covers much of the lunar surface (above). This photograph shows an Apollo 16 astronaut raking it for samples of rock – chipped angular fragments produced by several billion years of bombardment by asteroids and meteoroids. Several metres deep, the dust and debris layer – which is often called the regolith – may also contain microscopic beads of glass, produced by the solidification of fine sprays of impact-molten rock.

When the Apollo 12 lunar module (background, below) landed on the Ocean of Storms (Oceanus Procellarum) in November 1969, it did so within a short distance of the unmanned spacecraft Surveyor 3 (foreground) which had soft landed there two and a half years earlier.

MARS

Thanks to the cameras of two American Viking spacecraft, the red face of Mars is today as familiar to planetologists as that of the Man in the Moon. The Viking photograph on the left shows part of the western hemisphere of the planet. In the middle is a relatively dark Australia-sized patch called Mare Erythraeum. Cutting into its north-western flank is the eastern extent of the Valles Marineris canyon system. Twice the length and three times the depth of the Red Sea, it is one of the most prominent features of Martian geography, or areography as some will have it. Sandwiched between the southern rim of Mare Erythraeum and the receding shadow of the morning terminator is the circular impact basin Argyre Planitia with the crater Galle showing clearly in its mountainous eastern wall. The canyons run parallel to and just below the Martian equator, so much of the area shown lies in the southern hemisphere where it was late autumn when this picture was taken. In some of the craters to the north of the growing southern polar ice cap, patches of early morning frost or fog can be seen. North of the equator a hazy late spring morning is already several hours old.

Mars is on average one and a half times further from the sun than the earth. Its year – the time the planet takes to complete one circuit of the sun – lasts ten and a half earth months longer than ours. But its day, the Martian 'sol' – the time the planet takes to complete one turn about its axis – is only thirty-seven minutes more than twenty-four hours. The tilt of that axis is a mere

A photograph of the red planet taken by Viking 1 as it approached Mars in June 1976.

$\frac{1}{2}°$ greater than that of the earth's axis and similar to it in that the southern is the facing hemisphere when Mars is closest to the sun. And the difference between the perihelion and aphelion sun–Mars distance is 42,000,000 kilometres. So a Martian southern hemisphere summer is in practice what its earthly equivalent is only in theory – shorter and hotter than its northern counterpart.

Launched from earth orbit in August 1975, Viking 1 traced a 700-million-kilometre interplanetary spiral to reach Martian orbit the following June. In July 1976 a car-sized tripod lander, dropped from the mother craft, alighted on the rock-strewn surface of Chryse Planitia some 2,000 kilometres north of the eastern end of the Mariner canyon system. It was the first time a spacecraft had survived the landfall on Mars.

A dozen Julys earlier, the cameras of another American Martian probe, Mariner 4, returned twenty-two frames as it flew past the planet – the first of several thousand Mariner images from which a photographic atlas was made. In those days the feat of transmitting pictures – each of which consisted of 40,000 separate dots translated into bleeps – across 217 million kilometres of space was in itself a masterly achievement, but what they showed was less than cause for celebration. At the turn of the century the romantic canal maps charted by the American Percival Lowell ('That Mars is inhabited by beings of some sort or another we may consider certain,' he wrote, 'as it is uncertain what those beings may be.') had fired the imaginings of fellow astronomers and science fiction writers alike. And although by 1964 no one really expected to see

133

either these waterways or the Martian plantations which they were once supposed to have irrigated, the Mariner 4 photographs disappointed. Bleak and bare, a crater-pitted wilderness, Mars seemed utterly lifeless – not only biologically but geologically as well. Five years later, pictures from Mariners 6 and 7 (in July and August 1969) showed more of Mars and yet more craters. And at first an even less encouraging prospect greeted the lenses of a fourth spacecraft, Mariner 9, in November 1973 – a planet obscured by a global dust storm. But this Mariner was an orbiter, so time was on its side.

As the pink dust settled a new picture of the red planet began to emerge. Through the pale sinking clouds appeared four dark peaks – the summits of four enormous volcanoes, the largest of which, Olympus Mons, was found to rise from a 550-kilometre-diameter base to an altitude about three times greater than that of Mount Everest. Unveiled, Mars was now full of surprises which had eluded the earlier Mariners. For a small planet, the scale of Martian geology was impressive. Volcanoes, and canyons up to 240 kilometres wide and six and a half kilometres deep, pointed to an active past. Rocky plains, including that of Chryse Planitia, showed signs of having once been flooded. Sinuous valleys implied the flow of a fluid. Channels with branching drainage patterns strikingly similar to those of earthly gulleys formed by sudden rainstorms in desert regions suggested that this fluid might be water. The presence or past presence of water on Mars had earlier been discounted by all but ardent optimists among those planetologists who had studied the photographs from the first three Mariners. Now their faith seemed to have rewarded them. If the missing water could be found it was now reasonable to reopen the four-part question – could or can, did or does Mars support life?

Like that of Venus, the Martian atmosphere was known to be largely composed of carbon dioxide, although Mariner instruments showed it to be much thinner, with pressures on average one hundred times lower than those found on earth. Minute traces of water vapour detected in this tenuous air could hardly account for the dried-up river courses below. But in 1976 sensors aboard the orbiting section of Viking spacecraft revealed that while in the late northern spring it

was still very cold – minus 60 °C – the north pole of Mars was too warm to be composed of carbon dioxide. Those parts of the Martian polar caps which do not melt during the summer season must therefore be water ice. Another indication of the presence of water came from the orbiters' cameras. Some craters, it was noted, had ejected material splattered rather than scattered around their rims as though the colliding meteoroids which had formed them had landed in a sea of mud. The heat generated by such collisions, it was suggested, had melted permafrost. Just below the surface, perhaps, in many Martian regions there were vast reserves of water in this form.

Like those of the earth, both the orbit and axial tilt of Mars are known to vary in a cyclical fashion over periods of many thousands of years. And just as such variations in the distribution of sunlight around the planet may be a prime factor governing the waxing and waning of ice ages and other long-term weather patterns on earth, so they may bring about alterations of warm and cold ages on Mars. Long-term planetary climate cycles could also be caused by variations in the energy output of the sun, but whatever their cause it is reasonable to suggest that if such changes do occur, then Mars may currently be in the grip of a Martian ice age.

From the fourth ring of star-formed matter surrounding the sun, the ingredients of Mars

Looking north-east over the 1,000-kilometre-diameter Argyre Planitia basin in the Martian southern hemisphere, this black and white Viking 1 orbiter view shows the full sweep of the Charitum Montes chain which forms the southern half of its rim. Its eastern edge, marked by the large crater Galle, can be seen in the upper part of this photomosaic, which was made from photographs taken in July 1976, when it was late autumn at this Martian latitude. A thin layer of haze hangs over the horizon. The lower part of a Viking 2 orbiter photograph (colour inset) shows Charitum Montes seven earth months later. In February 1977 it was late winter in the Martian southern hemisphere. The Argyre basin lies near the margin of the south pole's winter extent and the mountains are capped with carbon dioxide snow. North of them, a dust storm (arrowed) blows eastwards across the basin.

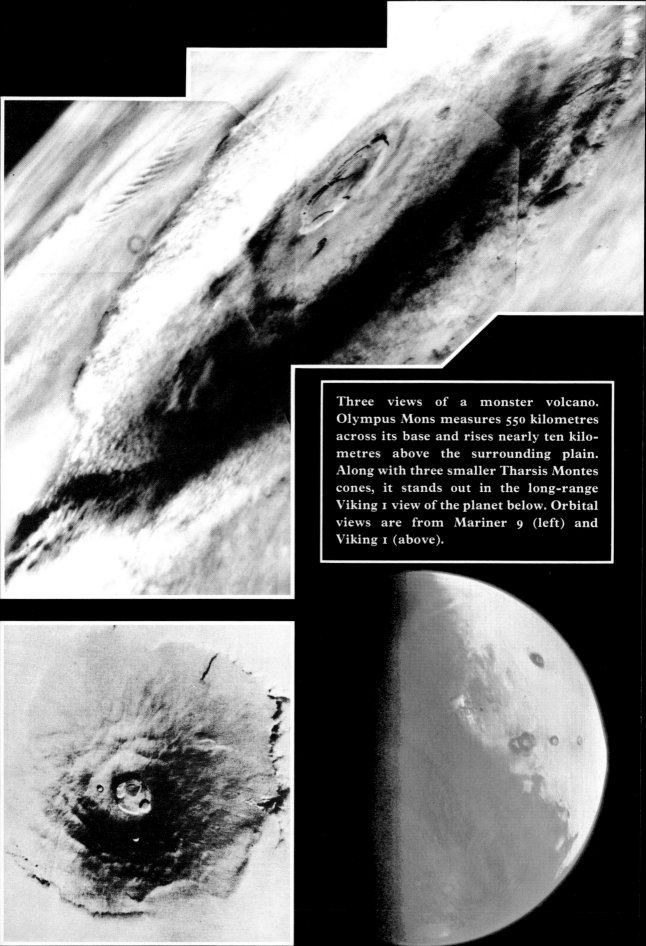

Three views of a monster volcano. Olympus Mons measures 550 kilometres across its base and rises nearly ten kilometres above the surrounding plain. Along with three smaller Tharsis Montes cones, it stands out in the long-range Viking I view of the planet below. Orbital views are from Mariner 9 (left) and Viking I (above).

were brought together by gravity some 4·6 billion years ago. And before it was reheated by the decay of its radioactive elements, the cooling planet acquired a crust. Then, as the interior became molten, heavier components sank to form a core while lighter fractions broke through to the surface to envelop the planet in a primordial atmospheric mixture of carbon dioxide, water vapour, methane, nitrogen and hydrogen. Martian gravity being insufficient to hold light-weight gases, the hydrogen soon evaporated into space but the density of that ancient atmosphere was certainly much greater than it is now.

Like that of Mercury and the lunar surface, the Martian crust is scarred by a number of colossal impact basins reflecting an early stage of solar system history when the orbits of the inner planets were still criss-crossed with those of substantial planetesimals. On Mars, a crater is reckoned to be ten to twenty times greater than that of the object which made it, so the body responsible for the 2,000-kilometre-diameter, and probably quite deep, basin of Hellas Planitia must have been of asteroid dimensions. The prominent circular features Isidis and Argyre Planitiae were also formed at about this time along with numerous smaller craters.

After a billion or so years the intensity of meteoroid bombardment subsided and new forces began to shape the Martian landscape. Vast pockets of lava erupted through the crust to cover vast areas of pitted craterscape with smooth volcanic plains. It is not unlikely that this face-lift was accompanied by a rejuvenation of the atmosphere. The scale of the lava flows suggests that concurrent outpourings of steam and gas may have made an appreciable difference to Martian atmospheric pressure. Following episodes of volcanic activity, the water vapour component of the atmosphere may well have become sufficient for rain to fall. Valleys cut into the older lava plains, dated by the presence of numerous sharp-rimmed younger craters, indicate the action of running water from around 3·5 billion years ago. Some are strikingly similar to those found on earth by run-off from rain- or spring-soaked uplands. Elsewhere they appear to have been excavated by swiftly moving bodies of meltwater thawed from the permafrost by underlying lava bodies rising beneath the surface. Because the Martian crust has long been

rigid – unlike the earth's, which is split into mobile plates – a Martian volcano, once formed, stays put immediately above its lava feeder pipes, increasing in size whenever fresh magma is supplied from below. Inactive Martian volcanoes can therefore be regarded as being dormant rather than extinct. The largest and probably the longest lived, Olympus Mons, presides over an extensive generally elevated region to the west of the head of the Mariner canyon system. It is possible that not all the material which raised this area came from below. Some of it may have been drawn from beneath the surface of adjacent regions. To the east, subsequent subsidence may well have been one of the first acts in the creation of the canyon system, although the melting and withdrawal of sub-surface ice and running water probably played their part.

From their interpretation of Mariner photographs, geologists helping to plan the Viking missions determined that the oldest, heavily cratered, lava fields must have poured across the surface more than two billion years ago and that subsequent phases of vulcanism have occurred throughout Martian history – the relative youth of the younger flows being indicated by their fresh, almost crater-free appearance. On Mars the gaseous components of even a small eruption would have a significant local if only temporary effect on the temperature and pressure of the planet's thin atmosphere. The effects of 'outgassing' associated with a sustained epoch of large-scale and planet-wide volcanic activity might well be global and lasting. Were such a major volcanic epoch to coincide with a period of global warming due to a favourable peak in the planet's axial or orbital cycle, or to a phase of enhanced energy output from the sun, there would, it has been suggested, be a dramatic transformation of the Martian environment. The atmosphere would become denser, enriched with carbon dioxide, water vapour and fine ash particles from volcanic vents, with dust whipped up by powerful winds generated by sudden pressure changes due to eruptions and by melting of permafrost and polar caps. Now an increasing part of the sun's heat reflected by the Martian surface would be absorbed by the atmosphere to be re-radiated back to the surface. Assuming that the ingredients and catalysts were present – volcanic gases and solar ultraviolet radiation for

Mars East

A geological map (right) shows the eastern hemisphere of Mars. It may be compared with an air brush drawing (below) which includes much of this hemisphere. The prominent dark plain (middle, left) is Syrtis Major Planitia (9; see also photograph, below left).

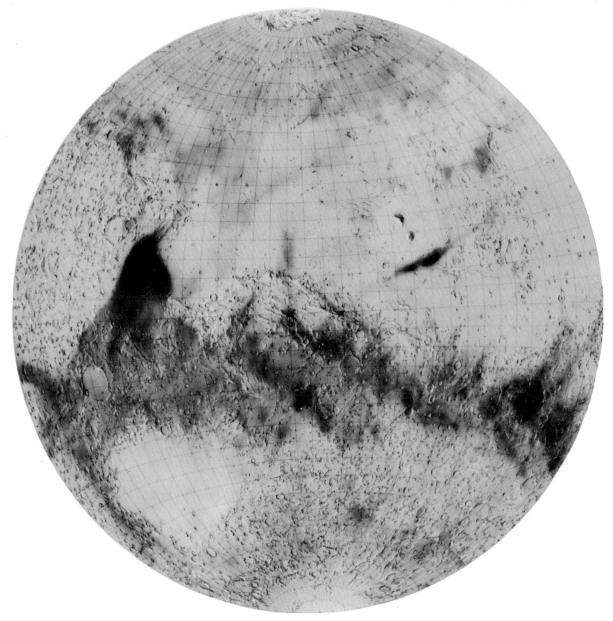

Sparsely to moderately cratered lowland plains like Vastitas Borealis (1 — see also key, right) occupy much of the northern hemisphere of Mars. Here their junction with the generally older and moderately to densely cratered uplands, which occupy much of the southern hemisphere, is marked by a low-lying zone of fretted hummocky terrain which includes isolated cliff-sided tablelands (12, 13).

The prominent dark equatorial feature, Syrtis Major Planitia (9 — see also photograph, left) is one of the most densely cratered plains regions of the planet. There are two similar zones north-east and south-west of Hellas Planitia (28).

To the east of Elysium Planitia (3), a sparsely cratered volcanic plain, is a region characterized by knobs, each of which is about ten kilometres in diameter.

The ice caps rest on layered terrain produced by alternating deposits of dust and ice. Some of the polar planes bordering the caps have wind-etched surfaces.

In the following list of formations shown on the geological map, the location of each formation is indicated by a key number followed by an indication of the quadrant in which it is found: north-west (NW), north-east (NE), south-east (SE) or south-west (SW).

Craters

Antoniadi	17	(NW)
Cassini	16	(NW)
Herschel	21	(SE)
Huygens	20	(SW)
Lyot	11	(NW)
Mie	15	(NE)
Schiaparelli	18	(SW)
Schroeter	19	(SW)
Secchi	32	(SE)

Fossae (long narrow valleys)

Nili Fossae	14	(NW)

Mensae (tablelands)

Deuteronilus Mensae	10	(NW)
Nilosyrtis Mensae	13	(NW)
Protonilus Mensae	12	(NW)

Montes (mountains and volcanoes)

Elysium Mons	5	(NE)
Hellespontus Montes	29	(SW)
Phlegra Montes	4	(NE)

Paterae (shallow craters)

Amphitrites Patera	30	(SW)
Apollinaris Patera	22	(SE)
Hadriaca Patera	27	(SE)
Tyrrhena Patera	26	(SE)

Planum (elevated plain)

Hesperia Planum	25	(SE)

Planitiae (plains)

Elysium Planitia	3	(NE)
Hellas Planitia	28	(SW)
Isidis Planitia	8	(NW)
Syrtis Major Planitia	9	(NW)
Utopia Planitia	2	(NE)

Tholi (hills)

Albor Tholus	6	(NE)
Australis Tholus	31	(SW)
Hecates Tholus	7	(NE)

North Pole

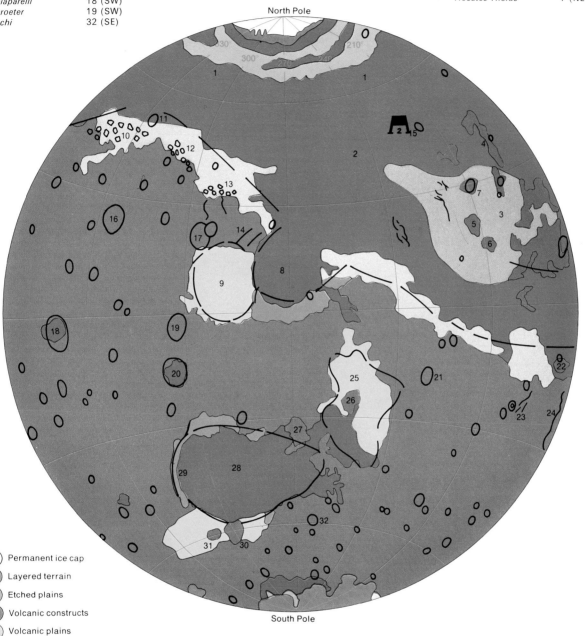

South Pole

○ Permanent ice cap

◐ Layered terrain

◐ Etched plains

◐ Volcanic constructs

○ Volcanic plains

○ Moderately cratered plains

○ Cratered plains

○ Fretted hummocky terrain

◐ Chaotic hummocky terrain

● Knobby hummocky terrain.

◐ Channel deposit

◐ Undivided plains

◐ Grooved terrain

● Undivided cratered terrain

○ Mountainous terrain

Valles (valleys)

Al Qahira Vallis	23	(SE)
Ma'dim Vallis	24	(SE)

Vastitas (widespread lowland)

Vastitas Borealis	1	(N)

 Viking mission number and landing site

Mars West

This Martian hemisphere (geological map below, air brush map right) is dominated by the Mariner canyons (18) and the volcano Olympus Mons (4), which, together with the three Tharsis Montes cones (5), also show up well in the Viking 2 photograph (below, left).

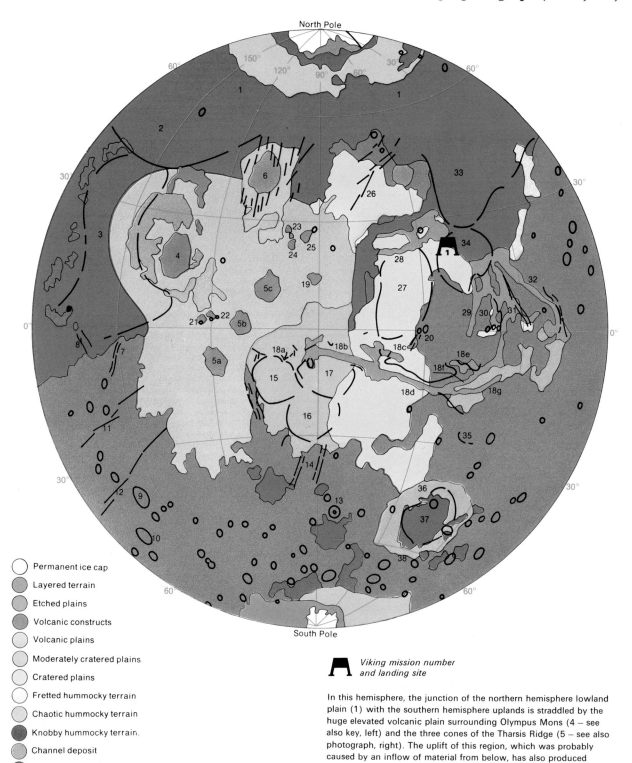

North Pole

South Pole

○ Permanent ice cap

● Layered terrain

● Etched plains

● Volcanic constructs

○ Volcanic plains

○ Moderately cratered plains

○ Cratered plains

○ Fretted hummocky terrain

● Chaotic hummocky terrain

● Knobby hummocky terrain.

● Channel deposit

● Undivided plains

○ Grooved terrain

● Undivided cratered terrain

○ Mountainous terrain

⋀ Viking mission number and landing site

In this hemisphere, the junction of the northern hemisphere lowland plain (1) with the southern hemisphere uplands is straddled by the huge elevated volcanic plain surrounding Olympus Mons (4 – see also key, left) and the three cones of the Tharsis Ridge (5 – see also photograph, right). The uplift of this region, which was probably caused by an inflow of material from below, has also produced numerous cracks in the surrounding surface.

These long narrow valleys (8, 11, etc.) are called fossae. West of this plain, the Mariner Valley canyon system (18) dominates the equatorial region north of the mountainous rim of the impact basin, Argyre Planitia (37).

In the following list of formations shown on the geological map, the location of each formation is indicated by a key number followed by an indication of the quadrant in which it is found: north-west (NW), north-east (NE), south-east (SE) or south-west (SW).

Chasmae (canyons)

Capri Chasma	18f	(SE)
Coprates Chasma	18d	(SE)
Eos Chasma	18g	(SE)
Gangis Chasma	18e	(SE)
Juventae Chasma	20	(SE)
Ophir Chasma	18c	(SE)
Tithonium Chasma	18b	(SE)
Valles Marineris	18b–g	(SE)

Craters

Copernicus	10	(SW)
Galle	39	(SE)
Lowell	13	(SE)
Newton	9	(SW)

Fossae (long narrow valleys)

Medusae Fossae	8	(SW)
Memnonia Fossae	11	(SW)
Sirenum Fossae	12	(SW)
Tempe Fossae	26	(NE)
Thaumasia Fossae	14	(SW)

Mare ('sea')

Mare Erythraeum	40	(SE)

Montes (mountains and volcanoes)

Arsia Mons	5a	(SW)
Ascraeus Mons	5c	(NW)
Charitum Montes	38	(SE)
Nereidum Montes	36	(SE)
Olympus Mons	4	(NW)
Pavonis Mons	5b	(NW)
Tharsis Montes	5a–c	(NW)

Labyrinthus (valley complex)

Noctis Labyrinthus	18a	(SW)

Paterae (shallow craters)

Alba Patera	6	(NW)
Biblis Patera	21	(NW)
Ulysses Patera	22	(NW)
Uranius Patera	25	(NW)

Plana (elevated plains)

Lunae Planum	27	(NE)
Sinai Planum	17	(SE)
Solis Planum	16	(SW)
Syria Planum	15	(SW)

Planitiae (plains)

Acidalia Planitia	33	(NE)
Amazonis Planitia	3	(NW)
Arcadia Planitia	2	(NW)
Argyre Planitia	37	(SE)
Chryse Planitia	34	(NE)

Tholi (hills)

Geraunius Tholus	24	(NW)
Tharsis Tholus	19	(NW)
Uranius Tholus	23	(NW)

Valles (valleys)

Ares Vallis	32	(NE)
Kasei Vallis	28	(NE)
Mangala Vallis	7	(SW)
Nirgal Vallis	35	(SE)
Shalbatana Vallis	29	(NE)
Tiu Vallis	31	(NE)
Valles Marineris	18b–g	(SE)

Vastitas (widespread lowland)

Vastitas Borealis	1	(N)

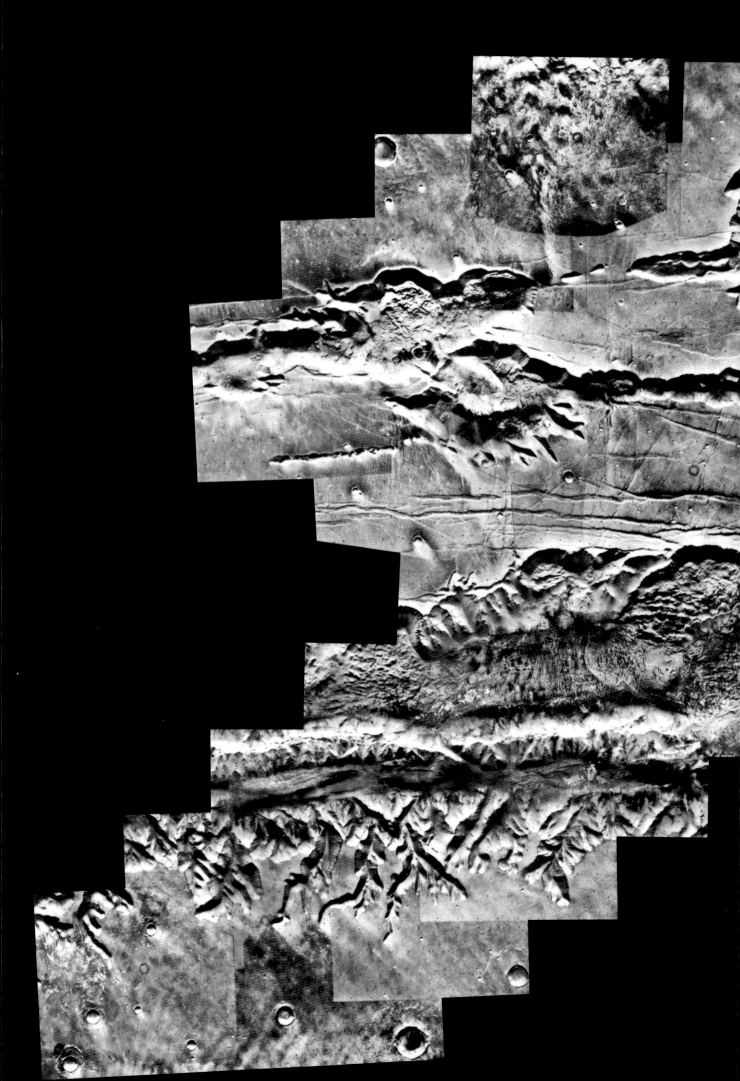

The painting overleaf shows a dust storm crossing a Martian canyon. These tempests are generated during the southern-hemisphere summer when the planet is closest to the sun. Like earthly rift valleys, the Mariner Valley canyons were probably formed by faulting. They are enlarged by wind erosion and landslides triggered by melting permafrost.

The mosaic above shows part of the 3,000-kilometre-long Valles Marineris or Mariner Valley canyon system which runs from west to east (from around longitude 90° to 40°) just south of the Martian equator. The lower of the two channels, Ius Chasma, is hereabouts around 60 kilometres wide and a thousand metres deep. To the north (above), one of a pair of comparatively narrow ravines joins the eastern end (left) of Tithonium Chasma with the western end (right) of Candor Chasma. The colour photograph shows the eastern end of Coprates Chasma where it meets the Capri–Eos junction.

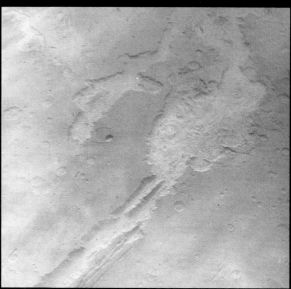

The photomosaic above shows slumping in a Mariner Valley canyon wall. The collapse of the wall has produced a series of huge land-slides which have spread across the canyon floor. Cutting into a crater on the plateau above, a large slide in the centre overlaps an older flow on the left. The older flow has two sharp-rimmed fresh craters in its left flank.

In the Viking 1 orbiter photomosaic below, a tear-drop-shaped feature on the left looks like an island in a river. Formed in its wake, the sharp end points downstream. Suggestive of a watery era in the Martian past, it has a series of concentric edges – a succession of beaches, perhaps, left by an age of drying climates and receding shorelines.

example – the question posed by biologists was did such 'greenhouse' conditions, which could account for past Martian rainstorms, ever persist long enough for the germination of Martian life? Today the atmosphere of Mars is more than one hundred times thinner than that of the earth. And while the planet's surface is cauterized by the steady blaze of solar ultraviolet, the temperature, except for a few of the daylight hours at equatorial latitudes, stays well below that of freezing water. So current Martian conditions must be considered hostile to life. But on earth evolution has enabled life to adapt to unfavourable circumstances – polar ice-dwelling bacteria for example. So – assuming that life had begun on Mars at a more favourable period in its history – had a radiation-proof, perhaps ice-consuming form evolved? Do thick-coated seeds lie dormant in the red Martian dust as their terrestrial equivalents do throughout the years of drought in the sands of earthly deserts?

Chryse Planitia, the 'golden plain', shows many signs of having once been flooded. Here was a good place, thought the Viking mission planners, to look for life or its remains. In July 1976 a small laboratory aboard the Viking 1 lander began its work. A mechanical arm reached out to scoop up samples of Martian soil and deposit them in a hopper atop the spacecraft whence they were distributed to different parts of an automated analyser no bigger than a small television receiver.

One of the experiments was designed to discover whether anything in the Martian soil takes up carbon from the air, as plants do on earth, to synthesize the organic compounds they need. So into the thimble-sized analyser chambers came soil samples to be mixed with fresh supplies of 'air' made up of two carbon gases known to be present in the Martian atmosphere, carbon monoxide and carbon dioxide. The carbon in the gases included the traceable radioactive variety, carbon 14. The samples could be warmed, watered and irradiated with ultraviolet-free light. At the end of an incubation period lasting five days the radioactive gases were flushed out. Then the samples were baked to convert any compounds the supposed Martian microbes had made back into gas. If any of this gas contained carbon 14, it could only have come from the Viking 'air' supply via something which

The slumped flanks of the Acidalia Planitia crater Arandas in the Viking 1 orbiter photograph, above, strongly suggests an impact in muddy ground. Another photograph, below, shows what appear to be flood-cut channels in the south-western region of Chryse Planitia. The biological significance of wet periods in the Martian past remains a mystery.

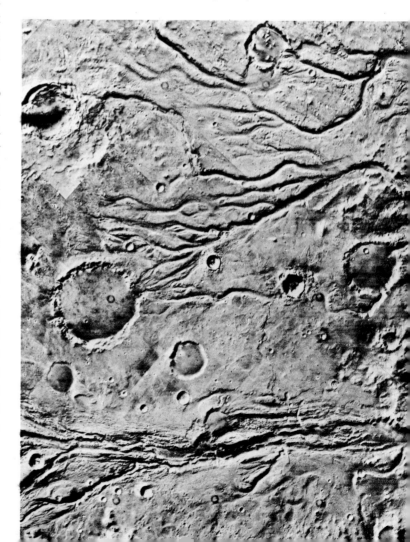

The landing section of Viking 1 alighted on the desert plain of Chryse Planitia in the northern hemisphere of Mars on 20 July 1976. The panorama above was photographed on 23 July and shows the landing site on a late spring afternoon. Projecting into the sky on the left are the struts of the lander's main antenna. The meteorology boom can be seen on the right.

The boom also cuts across the centre of a more detailed view of part of this scene (below). Behind the boom, loose rocks and sand dunes can be seen. Their alignment and sharp crests suggest strong winds blowing from the upper left of the picture.

Some idea of the scale of the scene may be judged from the fact that the large rock on the left measures about three metres across and lies about eight metres from the spacecraft. The horizon is about three kilometres distant.

The hills behind the meteorology boom may form part of the rim of a crater.

Colour photographs of parts of the site, obtained by scanning the same view three times through different filters, show the rusty orange hue of the Chryse Planitia wilderness. The colour is probably due to an abundance of iron oxides in the surface materials. If this is so, and if they were formed on Mars in the same way as they are on earth – by weathering of rocks in the presence of oxygen and water – these reddish minerals suggest a past Martian era when oxygen and water were more plentiful than they are at present. The sky's pinkish tinge is thought to be caused by fine dust suspended in the atmosphere.

The rock-strewn area on the right closely resembles the surface of a stony desert on earth. This is the location of mankind's first on-site search for life on another planet.

This panorama shows the rock-strewn surface of the Viking 2 landing site on Utopia Planitia about 200 kilometres south-west of the crater Mie. What seems to be a sinuous channel winds its way from the upper left to the lower right. Two colour photographs (below) show seasonal changes. The picture on the left was taken in the Martian northern hemisphere summer (November 1976). On the right, part of the same scene is shown in winter (October 1977). Note frost patches.

had acquired it during the incubation period.

Suitably encoded for transmission from Chryse Planitia to Pasadena, California, the carbon 14 measurements from several runs of this experiment produced a subdued surge of optimism when they reached the Viking control centre. Something in the Martian soil *was* taking up radioactive gas and, like earthly microbes, was largely destroyed by heating. But the reaction was immediate, rapid and intense – quite unlike an earthly biological process. Similar results were reported from the same experiment carried out by Viking 2 which landed on Utopia Planitia, a third of the way round the planet to the east of Chryse Planitia, in September 1976. The 'something' seemed to take up gas more readily when exposed to light but was apparently greatly inhibited by the presence of liquid water, either when it was intentionally introduced during the experiment or when it melted from trapped ice particles in a soil sample taken from under a rock.

In a second experiment a soil sample was 'fed' a few drops of nutrient in the presence of the normal Martian atmosphere to see whether anything in the soil would consume this food, releasing carbon dioxide in the process as earthly organisms do. Once again the carbon components of the added material included the radioactive variety. Any carbon dioxide released by decomposition of the nutrient would be similarly 'labelled'. Data from the first run of this experiment on Chryse Planitia indicated an initial surge of gas production. Clearly something was 'eating' the food. On the seventh day a second course of the same nutrient was served. This caused first a second surge but then a sharp drop in the level of radioactive gas present. When a second sample was sterilized by heating before it was 'fed' there was almost no gas production. And in a similar run of this experiment aboard Viking 2, heating to only 50 °C disabled whatever was decomposing the nutrient.

For the third experiment samples of soil were sealed in an analyser chamber connected with a gas chromatograph which could detect any change in the atmospheric mix following the addition of water or nutrient. Samples from both sites produced significant results. Not only did apparent consumption of the nutrient give rise to a steady rate of carbon dioxide production but

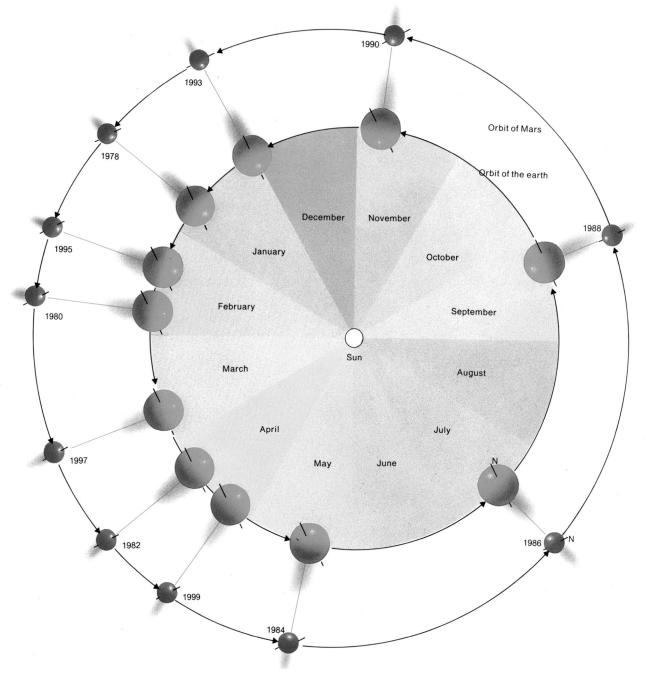

Orbit of Mars

Orbit of the earth

Sun

1990
1993
1978
1995
1980
1997
1982
1999
1984
1986
1988

December
November
October
September
August
July
June
May
April
March
February
January

N
N

the introduction of water vapour resulted in a most unexpected surge in oxygen levels. Was this perhaps the awakening of some long dormant Martian seed or spore?

On earth living systems are constructed from organic molecules most of which consist of varying amounts and proportions of carbon and hydrogen together with oxygen, nitrogen and other elements. It is assumed that the same would hold true for any Martian organisms. In a sample of earth soil the volume of organic waste products and decomposed remains greatly exceeds that of living cells themselves. Thus even if the Martian microbes were too sparse to be detectable, the presence of organic compounds in a soil sample would be indicative of their existence. But at neither site did Viking detectors find any organic molecules. Even those which had presumably arrived as components of meteorites had apparently been destroyed by solar ultraviolet radiation at the Martian surface. Yet, taken together, the results of the Viking experiments were, in the words of biology team leader Harold Klein, 'somewhat controversial'. On the one hand something in the Martian soil had consumed carbon dioxide, 'eaten' nutrient and produced oxygen on contact with water vapour, all of which could be regarded as indicative of biological processes. On the other hand there were no bodies, nor any of their remains. The tentative answer to this enigma had to be the conservative one − that the exotic chemical behaviour of surface samples of Martial soil was an inorganic not a biochemical phenomenon.

While to some this was a disappointing conclusion, it could not yet be said that the quest for life on Mars had been in vain. Although the two locations tested were almost certainly barren, it had been impossible to extend the search beyond a few centimetres below the exposed surface. Martian organisms, it could still be argued, might yet be found at greater depth or at sites offering

a measure of protection from the full force of solar ultraviolet radiation. Nor did the Viking results rule out the prospect of finding the similarly preserved fossil remnants of more flourishing times.

As well as looking for signs of life, Viking landers were equipped to measure other properties of the planet. A Marsquake detector aboard Viking 2 revealed something of its interior. Although scanty the data indicated a crustal thickness of about 15 kilometres at the Utopia site. Elsewhere it may prove thicker. Viking analyses indicate that the Martian crust is made up of much the same components and in much the same proportions as the earth's. But, taken as a whole, Mars is only two thirds as dense as our planet. Its core, therefore, is probably correspondingly less metallic than the earth's. It is our planet's iron-rich core which gives the earth its high density − and its magnetic field. Mars has no significant magnetic field although iron is present. On the surface it is in a form which gives the planet its reddish hue.

The spacecraft were also fitted out as weather stations. Summer temperatures at both sites, 22° and 48° north of the equator, varied between a pre-dawn minimum of below minus 80 °C and a sunny minus 30 °C. Light winds were recorded during the northern Martian spring. During the same season the pressure at both sites, initially 7·7 millibars at Chryse and 7·74 at Utopia, showed a small but persistent daily decrease before levelling off in the northern summer at around 6·5 millibars and 7·4 millibars respectively. This was thought to be due to the removal of atmospheric carbon dioxide by the autumnal and growing southern ice cap at a more rapid rate than its springtime release by its nothern counterpart. As the northern hemisphere summer drew to a close winds became stronger and dust storms were seen in the photographs returned from Martian orbit. Such seasonal shifts of large bodies of dust are very likely the cause of colour changes once said to be indicative of the growth of vegetation.

Named after the charioteer sons of the ancient Greek war god Ares, the two moons of Mars (Ares in his Roman guise) Phobos (fear) and Deimos (terror) are very small and very dark. Orbiting their parent much more closely than does our moon, they are difficult to see from earth

The diagram (above left) shows the calendar of Mars−earth close approaches from 1978 until the end of the century. Mars-bound spacecraft are likely to be launched some months before, to arrive some months after, these dates. The photograph shows part of the Nilosyrtis Mensae region to the west of Utopia Planitia where freeze−thaw cycles may have governed seasonal erosion by mud flows.

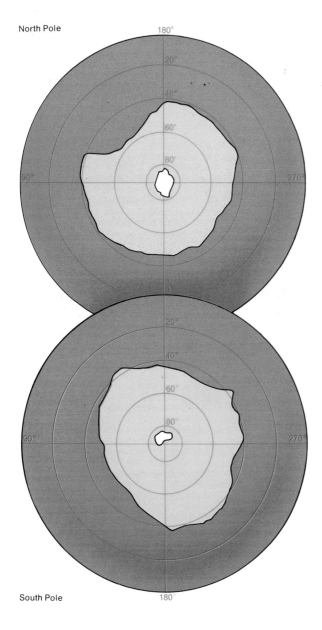

North Pole

180°

20°

40°

60°

80°

90°

270°

South Pole

180°

The colour photograph on the right, taken by the Viking 2 orbiter in October 1976, shows an edge of the north polar ice cap in the Martian summer. The mantle of carbon dioxide ice which overlies and considerably extends the pole in winter has lifted to reveal a series of fifty-metre-high steps cut into the layered flanks of a 500-metre-high scarp bordering the southern rim of the ice cap in the upper part of the photograph. The layering is probably due to alternating blanket deposits of dust and ice.

Part of the summer core of water ice can also be seen in the black and white photograph. The dark ice-free bands may have been formed by polar winds.

The diagram above compares the winter (white) and summer (pale blue) extents of the two Martian poles.

154

due to earth-scattered light from Mars and remained undiscovered long after the invention of the telescope. Not until an exceptional instrument, the United States Naval Observatory 16-inch refractor, came into the hands of an exceptional observer, Asaph Hall, were they found, in 1877.

Potato-shaped, both satellites are aligned with their long axes directed towards the planet. That of Phobos measures 27 kilometres. The vertical component, from north pole to south pole, is 19 kilometres while the least equatorial width is 21 kilometres. Corresponding dimensions for Deimos are 15, 11 and 12 kilometres. The nearly perfectly circular and equatorial 9,350-kilometre orbit of the larger moon – measured across the Martian equatorial plane from the centre of the planet – may be compared with those of artificial earth satellites. A sidereal Phobos month is very brief. Rising in the west and setting in the east, Phobos takes a mere 7 hours 42 minutes to circle the planet and lies within the 'Roche limit' where disruptive Martian gravitational effects, it has been suggested, may eventually pull the moon apart to make Mars the third ringed planet of the solar system.

At 23,490 kilometres from the centre of the planet, the circular equatorial orbit of the smaller outer moon is just beyond that of a 'stationary' satellite when orbital velocity matches the rotation of the Martian landscape below. Deimos takes 30 hours 18 minutes to circle Mars. Like ours, both moons keep one hemisphere pointed towards the parent planet.

Photographs taken by Mariner 9 and Viking spacecraft show both moons to be pitted with craters old and new. And in spite of their very low escape velocity, such that many of the fragments dislodged by an impacting meteoroid would fly off into space, both moons appear to be coated with a thick layer of fine black powder. Remarkable similarities in shape and in orbit suggest that Phobos and Deimos share a common origin but whether they were formed from the same material and at the same time as Mars, or were acquired later by capture from the nearby asteroid belt remains uncertain. What is certain, and perhaps disappointing, is that Phobos is not the hollow artificial product of Martian engineering once suggested by the distinguished Russian astronomer Iosef Shklovsky.

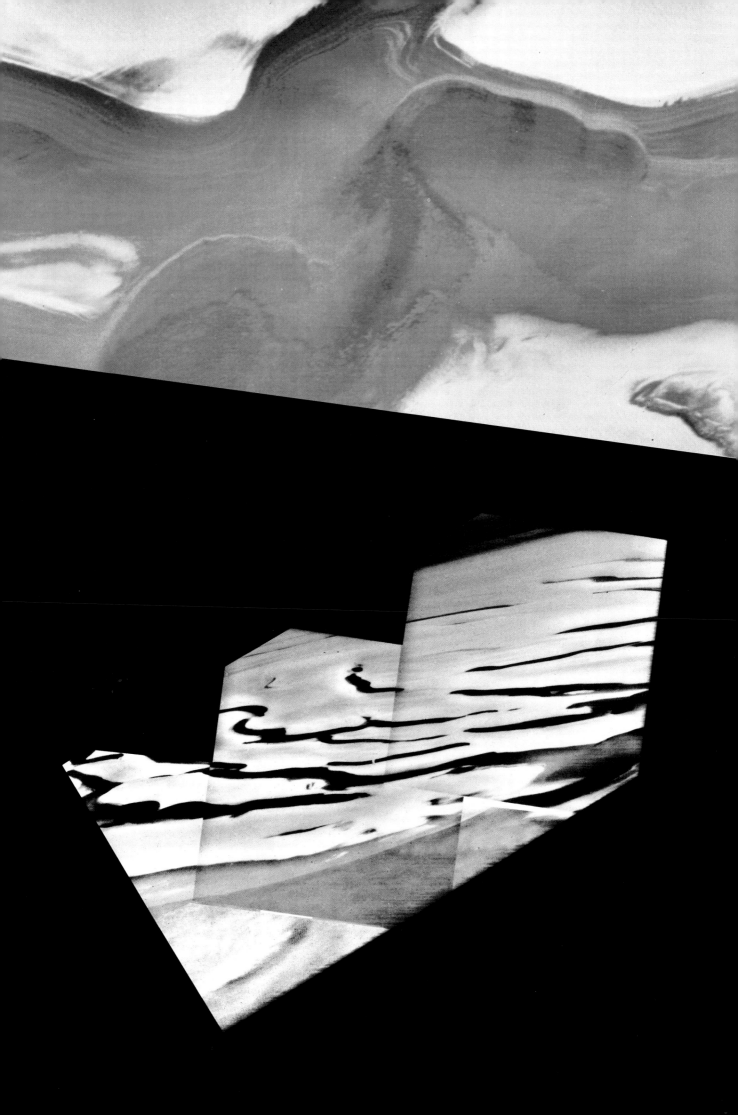

The photograph above shows what looks like a ploughed field. It was taken by the Viking 1 orbiter at longitude 350°, latitude 46° north. Possibly formed by wind erosion, the parallel valleys and ridges are about one kilometre apart. Another Viking 1 orbiter photograph (below) shows the potato-shaped silhouette of the Martian moon Phobos passing beneath the spacecraft as it overflew an area south-east of Chryse Planitia. The painting on the right is a romantic version of what many people hoped the Viking landers' cameras would show.

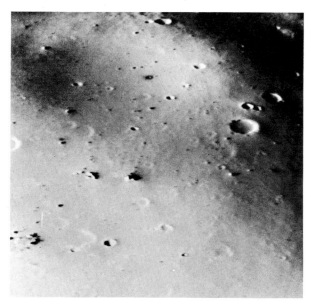

The painting on the left shows the planet setting behind Deimos, the outer and smaller of the two Martian moons. Deimos is also seen in a Viking 1 orbiter image (above right). Its dimensions are about 15 × 12 × 11 kilometres. A close-up photograph of the moon (above) shows small craters almost completely buried in a layer of dust.

A mosaic of Viking 1 orbiter photographs (below right) shows the inner and larger moon, Phobos. Its dimensions are about 27 × 21 × 19 kilometres. North is at the top. The south pole lies within the largest (five kilometres-diameter) crater, named Hall after the moons' discoverer, in the bottom half of the picture. Another image (below) shows striations running parallel to the moon's equator and orbital plane. Their origin is uncertain but may possibly be due to Martian tidal forces which are stretching the moon. Being so close to the planet, Phobos is under considerable strain and may eventually disintegrate to form rings around Mars.

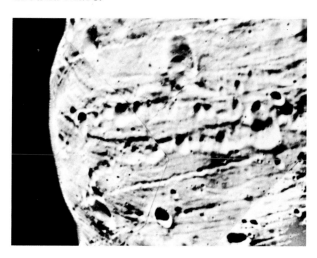

ASTEROIDS

Johann Bode, the German astronomer, is best remembered for his rule of thumb known as 'Bode's Law' although it is neither a law in the strictest scientific sense nor was it invented by Bode. But it was he who promoted the discovery, made in 1772 by another German astronomer Johann Titius, that if one adds 4 to each term in the numerical series 0, 3, 6, 12, 24, 48, 96 and then divides each sum by 10, the result is a strikingly good likeness to the list of average distances from the sun of the then known planets expressed in astronomical units (where the earth's distance is taken as unity). The figures obtained by using this formula compared with the true values, given in brackets, are – Mercury 0·4 (0·39), Venus 0·7 (0·72), Earth 1·0 (1·0), Mars 1·6 (1·52), 2·8, Jupiter 5·2 (5·20) and Saturn 10·0 (9·54).

The next term in the series is 19·6, so when the Hanoverian astronomer Sir William Herschel found the seventh planet, Uranus, in 1781 and its average distance from the sun turned out to be 19·18 astronomical units, Bode was reasonably well pleased. There was, it seemed, only one flaw. There was no known planet at the distance corresponding to the fifth term in the series, 2·8 astronomical units. But that, believed a Baron Franz von Zach, was only because it had yet to be found, and to this end he recruited a team of twenty-four fellow astronomers. It was not however von Zach and his so-called 'celestial police' but a professor at Palermo who did the

finding. On 1 January 1801, the first day of the nineteenth century, Giuseppe Piazzi discovered a star-like object, an asteroid, in an orbit between those of Mars and Jupiter. Named Ceres, it lies at an average distance of 2·77 astronomical units from the sun and is thus well qualified to fill the vacancy in Bode's line-up.

This elegant cosmic geometry was soon disturbed by the revelation that Ceres (estimated diameter 1,000 kilometres) has sisters. Today it is thought that there are perhaps 50,000 asteroids within the visual range of optical telescopes.

Most known asteroids lie within the 'main belt' between 2·2 and 3·3 AU from the sun. But others, such as the 'earth grazers', have elliptical orbits. Another group, the Trojans, lies in Jupiter's orbit, 60° ahead of and 60° behind the planet. And in November 1977 Charles Kowal found Chiron (estimated diameter 160 kilometres) in an orbit beyond that of Jupiter. A number of gaps in this band, caused by the considerable gravitational influence of Jupiter, correspond to orbital distances at which periods (the time taken to complete an orbit of the sun) are simple fractions of a Jovian year. Repeated close approaches of the giant planet have prevented asteroids from settling down in such orbits.

The larger asteroids may have metal-rich cores. Others may be rocky or metallic fragments – the products of collisions between the members of an earlier generation of perhaps thirty asteroids.

Many asteroids are very dark bodies. These, it has been suggested, may be covered with a layer of the most ancient material in the solar system, formed when the gas and dust cloud surrounding the young sun first began to cool.

Responsible for inner planet craters, asteroids were once more numerous. Most of the known survivors lie in a belt between Mars and Jupiter.

JUPITER

Embracing 71 per cent of the total planetary mass of the solar system, Jupiter lies between 4·95 and 5·45 astronomical units from the sun. Because the planet is so huge and its gravitational pull is so strong, little or none of the gas and dust mixture that went into its making, some 4·6 billion years ago, has since leaked away into space. Unlike the smaller planets, Jupiter retains an enormous proportion of hydrogen and helium in its composition (average density 1·3) and is therefore, except for its core, like the sun, a fluid rather than a solid body.

Deducing the internal form of a 142,800-kilometre-diameter planet from a distance never less than 589,000,000 kilometres is necessarily a highly speculative venture but, aided by data returned by two American spacecraft, Pioneers 10 and 11, which flew past Jupiter in December 1973 and December 1974, astronomers have been able to construct a tentative model of the giant's interior structure. At its core, it is suggested, lies an approximately earth-sized kernel of iron-rich silicates surrounded by a 93,000-kilometre-diameter sphere of liquid hydrogen so dense that the element is in the form of dissociated atoms rather than molecules. Under these conditions hydrogen behaves like a metal and convection currents in this layer are the probable cause of Jupiter's strong magnetic field. Enclosing this envelope of fluid metallic hydrogen is a 24,000-kilometre-deep outer 'ocean' of 'normal' liquid hydrogen.

Above the liquid molecular ocean the 1,000-kilometre-thick Jovian atmosphere is layered with water droplets and ice crystals at its base, crystals of ammonia and its compounds in the middle and gaseous hydrogen at the top. Like that of the earth, the Jovian atmosphere is divided into a series of circumplanetary convection cells lying parallel to the equator. Pure white to pale yellow *zones* of rising gas alternate with brownish down-draught *belts*. North and south of the equatorial zone run sequences of equatorial belts, tropical zones, temperature belts and temperature zones giving the planet its banded appearance. Generated by the planet's rapid rotation rate – a Jovian day, the shortest in the solar system, lasts a mere 9 hours, 50 minutes – strong Coriolis forces deflect north- and south-bound convective currents to give rise to powerful easterly and westerly winds. Consequent eddying between pale zones and dark belts produces all manner of multi-coloured swirls, loops and plumes.

Contrasting sharply with its biscuit-coloured surroundings among the cloud tops of the south tropical zone is the planet's most distinctive feature, the Great Red Spot. Forty thousand kilometres long and 11,000 kilometres wide, it was noted by the English astronomer Robert

Thirteen hundred times the volume of and three hundred times more massive than the earth, Jupiter is the solar system's largest planet. Turbulent circumplanetary atmospheric bands, laced with brightly coloured compounds of ammonia and perhaps sulphur, make it a spectacular object to view by telescope; this picture was taken by an American spacecraft, Pioneer 11, as it flew past the planet in December 1974. Raging for hundreds or perhaps thousands of years, a 175,000,000-square-kilometre storm produces the famous Red Spot in the southern hemisphere.

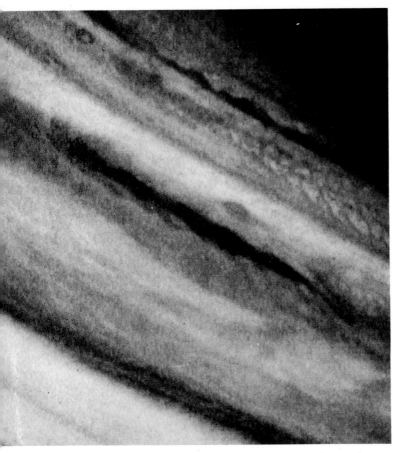

In the middle of the Pioneer 10 colour photo-graph (far right), taken in December 1973, a 65,000-kilometre streamer can be seen trailing eastwards from a bright nucleus in the equatorial zone. The nucleus is apparently being fed by a thermal beneath it. North of the equator, the junctions between bands become progressively more complex, reflecting turbulent weather conditions. Nearer the equator (detail above), the boundaries between zones and belts are more clearly defined.

Hooke in 1664, so it is at least 300 years old. But over the years it has varied in extent, hue and longitude, all of which indicates not a permanent feature but a long-lived meteorological event. Spiralling, seen in some of the Pioneer photographs, suggests that the Red Spot lies at the centre of a monster Jovian tropical storm.

Centrifugal distortion, another consequence of Jupiter's rapid rate of spin, gives the planet a somewhat flattened form. Its equatorial diameter is 8,800 kilometres greater than its polar diameter. Like another cloudy planet, Venus, the axis about which Jupiter rotates is only very slightly tilted with respect to the plane of the planet's orbit around the sun. And although the sun–Jupiter distance varies by as much as 74,800,000 kilometres throughout the Jovian year (11·86 earth years), it is almost always more than ten times that figure, so neither the planet's axial tilt nor the eccentricity of its orbit causes any significant variation in the total amount of radiation received from the sun. In any case temperatures on Jupiter depend more on the planet's endemic internal heat sources than on the remote star about which it revolves. Trapped since the formation of the planet or generated by the decay of its radioactive components, the heat energy contained in the Jovian core has driven up the temperature to an estimated 30,000 °K which is five times greater than that of the surface of the sun. On balance, Jupiter emits twice as much radiation as it receives. Thus, in some ways, the planet can be thought of as a 'failed' star. Largely composed of hydrogen, were it five or ten times more massive, internal temperatures and pressures might well be sufficient to create a stellar furnace.

Between the hot interior and the cold (minus 150 °C) cloud tops of the zonal bands, Jupiter offers a considerable range of temperatures, pressures and perhaps chemistries about which very little is yet known. In addition to hydrogen, helium, water and ammonia, the visible part of the Jovian atmosphere is known to contain methane which includes carbon in its make-up

Two Pioneer 11 photographs (left and right) show cloud tops near the north pole of Jupiter – a region which cannot be seen from earth. Where they meet, the edges of the atmospheric zones and belts are moulded into swirls and scallops by competing winds.

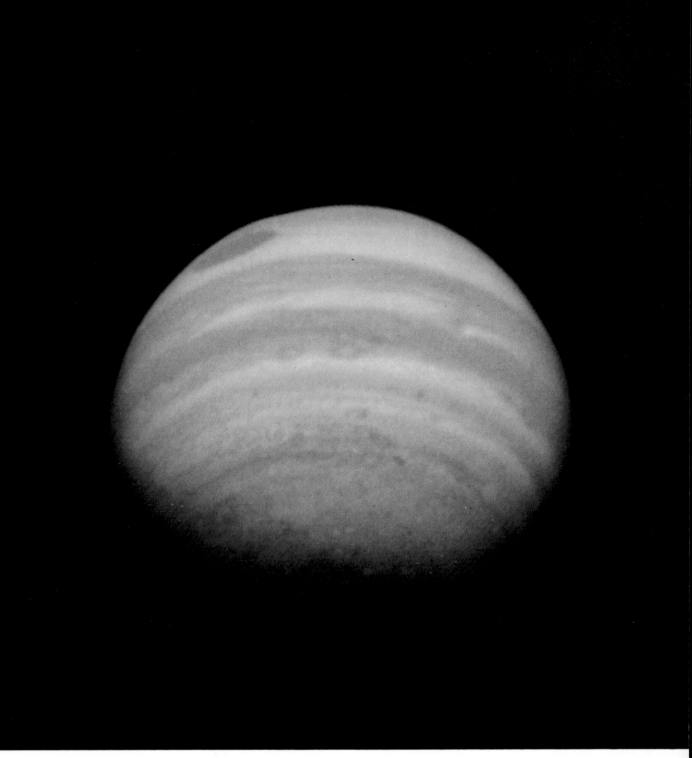

Taken as Pioneer 11 flew over the top of Jupiter in December 1974, the photograph above shows the planet from its north pole (top of image) to the Red Spot south of the equator.

The Pioneer 11 photographs on the right show the Red Spot in colour and by blue light alone (black and white inset). Structures within the Spot (which show up best in the blue-light image) suggest an anti-clockwise spiral motion, meshing with a right-to-left flow of white clouds north of the Spot and a left-to-right stream south of it. On the perimeter, triangular regions, west and east of the Spot,

are bounded by these two atmospheric masses flowing in opposite directions. The white oval, beneath the eastern end of the Spot, is one of three such ovals which travel round the planet with the atmospheric bands. Its circular central eye suggests rotation. A streamer of brownish material extends from the right of the Red Spot and the oval.

The painting overleaf shows the Red Spot viewed end on through a storm in the foreground. In the background, either side of a thermal, the sun (left) and a moon (right) can be seen.

166

and possibly sulphur which compounds with ammonia to produce colourful sulphides and hydrosulphides, which might explain the pigmentation of the belts and the Red Spot. The likely presence of such biologically significant compounds at lower atmospheric levels where temperatures are thought to be in the 35 °C to 75 °C range raises the intriguing question of Jovian life. But until we know more about the planet's internal heat source and its atmospheric chemistry we can only guess at what organism could prosper in its turbulent atmosphere.

Thanks to its massive internal shell of 'metallic' hydrogen, Jupiter has a strong magnetic field. Like that of the earth, this has produced an onion-structured external magnetosphere of trapped particles enclosed behind the bow of an interplanetary shock wave which rides the solar wind. Compared with its terrestrial equivalent, Jupiter's field is reversed so that its north magnetic pole is close to its south geographical pole. As well as radiating heat from its surface, Jupiter is a transmitter at radio frequencies. While some of these emissions originate in the Jovian atmosphere, others appear to come from the magnetic zone. The latter are noticeably affected by the presence, within the magnetosphere, of five of Jupiter's moons.

There are at least thirteen Jovian moons. Four of them are large with diameters greater than 3,000 kilometres. Three of these are bigger than the earth's moon. With diameters averaging only about 50 kilometres, the nine others are comparatively small. The moons can be divided, by orbit, into three lunar systems. The first, inner system consists of the four large moons and one small one. Their orbits are circular, concentric and lie within the magnetosphere in the same plane as Jupiter's equator. The second, middle system has four members. Tilted an average 28° with respect to Jupiter's equatorial plane, their orbits intertwine at an average distance of 11,500,000 kilometres from the centre of the planet. Another group of four moons with interlocking orbits makes up the third, outer system. At an average distance of 22,700,000 kilometres from the centre of the planet, their orbits are tilted an average $26\frac{1}{4}°$ with respect to Jupiter's equatorial plane. The motion of these moons is retrograde. Viewed from above the north pole of Jupiter, they journey clockwise round the planet.

Jupiter has at least thirteen moons. (At the time of writing, the discovery of a fourteenth awaited confirmation.) The shadow in the Pioneer 10 photograph above is that of Io, the innermost of four large Jovian moons, three of which (including Io) are bigger than the earth's. A computer plot on the right and a diagram overleaf shows their orbits.

Five concentric rings mark the circular equatorial orbits of the orderly inner system – the small moon Amalthea and the four giants, Io, Ganymede, Europa and Callisto.

Around the inner system are the eccentric interwoven orbits of the four middle-system moons, Leda, Himalia, Lysithea and Elara. None of them has a diameter greater than 100 kilometres.

The orbits of the four tiny (20-kilometres-across) outer-system moons, Ananke, Carme, Pasiphae and Sinope, are also eccentric, interwoven and unusual in that the moons journey round the planet in a clockwise or retrograde direction (unlike all planets and most other moons which travel anti-clockwise when viewed from above the plane of the earth's orbit around the sun).

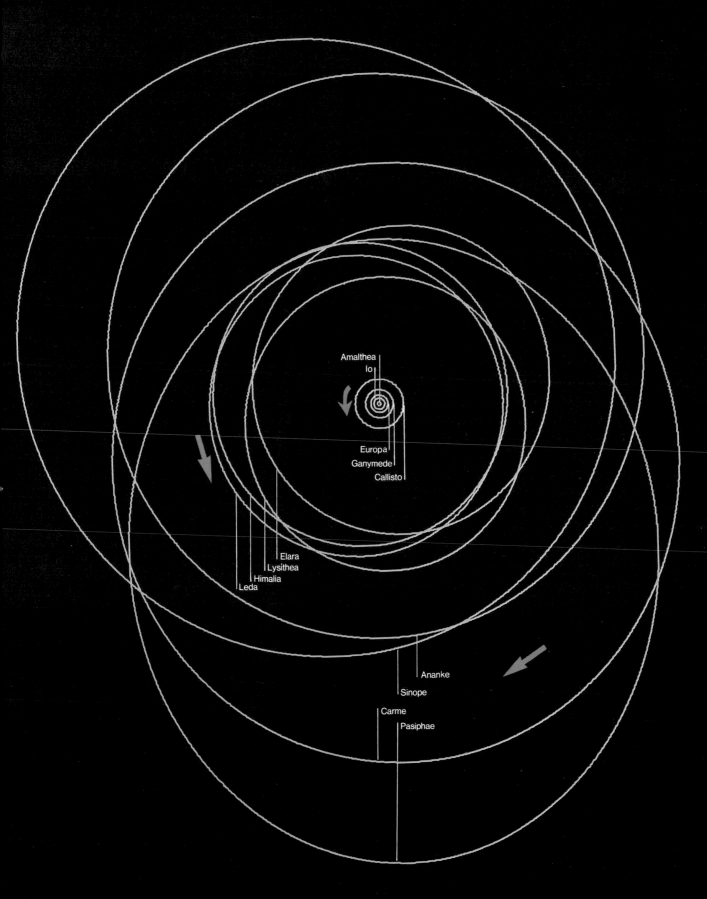

Amalthea
Io
Europa
Ganymede
Callisto
Elara
Lysithea
Himalia
Leda
Ananke
Sinope
Carme
Pasiphae

Approximate scale in millions of kilometres

0   5   10

Jupiter's moons

This diagram gives an oblique view of Jupiter's three lunar systems. The inner system moons have circular equatorial orbits. Those of the middle system have interweaving eccentric orbits, inclined an average of 28° to the plane of the planet's equator. Those of the outer system have interweaving eccentric retrograde orbits with an average inclination of $26\frac{1}{4}°$.

In the table below, diameters and average distances from Jupiter are given in kilometres. Eccentricity is a measure of how well a moon's orbit matches a circle (○). Orbital periods are relative to the stars.

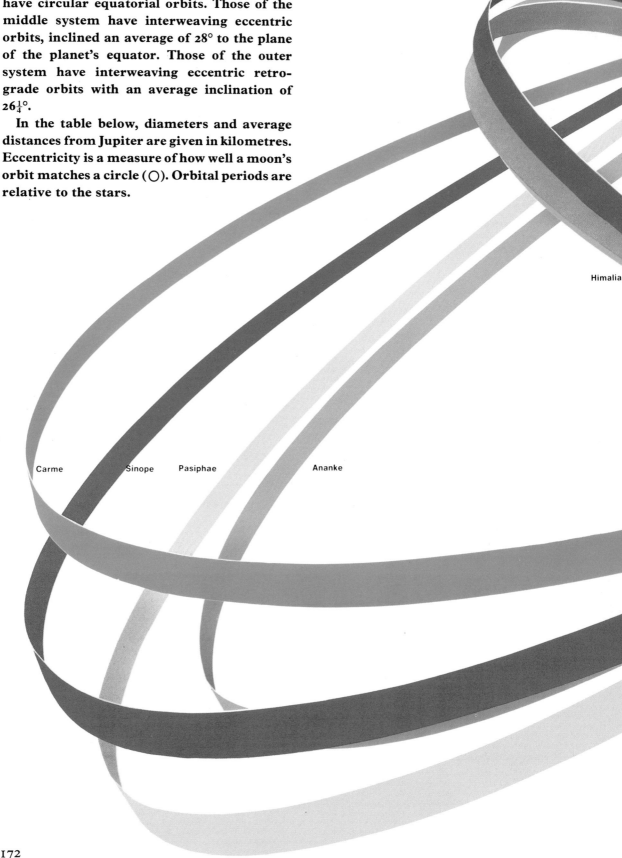

Himalia

Carme   Sinope   Pasiphae   Ananke

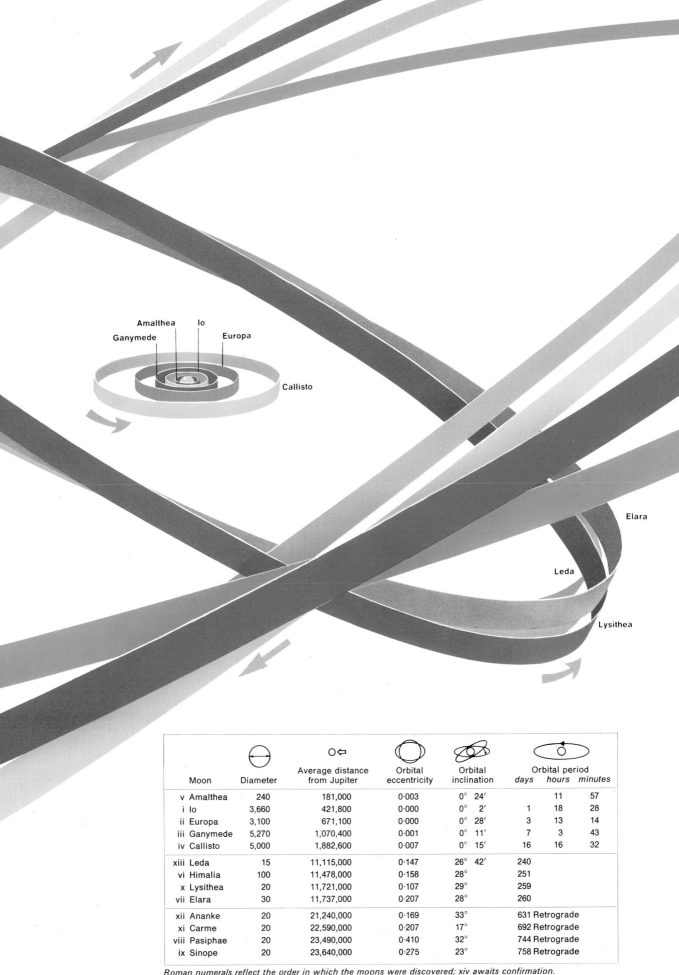

Ganymede Amalthea Io Europa

Callisto

Elara

Leda

Lysithea

Moon	Diameter	Average distance from Jupiter	Orbital eccentricity	Orbital inclination	Orbital period		
					days	hours	minutes
v Amalthea	240	181,000	0·003	0° 24′		11	57
i Io	3,660	421,800	0·000	0° 2′	1	18	28
ii Europa	3,100	671,100	0·000	0° 28′	3	13	14
iii Ganymede	5,270	1,070,400	0·001	0° 11′	7	3	43
iv Callisto	5,000	1,882,600	0·007	0° 15′	16	16	32
xiii Leda	15	11,115,000	0·147	26° 42′	240		
vi Himalia	100	11,478,000	0·158	28°	251		
x Lysithea	20	11,721,000	0·107	29°	259		
vii Elara	30	11,737,000	0·207	28°	260		
xii Ananke	20	21,240,000	0·169	33°	631 Retrograde		
xi Carme	20	22,590,000	0·207	17°	692 Retrograde		
viii Pasiphae	20	23,490,000	0·410	32°	744 Retrograde		
ix Sinope	20	23,640,000	0·275	23°	758 Retrograde		

Roman numerals reflect the order in which the moons were discovered; xiv awaits confirmation.

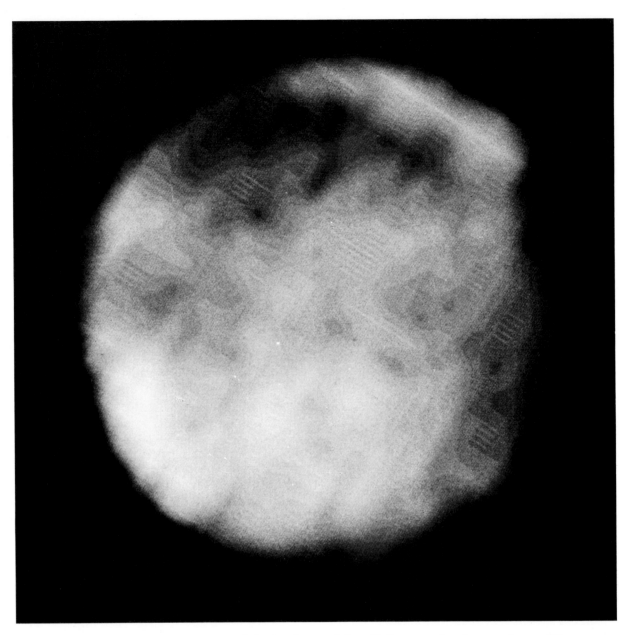

Ganymede (above) is the largest of Jupiter's moons. With a diameter of 5,270 kilometres, it is the third largest moon (after Neptune's Triton and Saturn's Titan) in the solar system. In this Pioneer 10 photograph, taken from a distance of about 750,000 kilometres, a dark region in the northern hemisphere contrasts with a light zone in the southern. Pioneer 11 photographs below show Ganymede (left), Callisto (centre) and Io (right). Much better images of these and other moons are expected when two American Voyager craft, launched from earth in the summer of 1977, reach the planet in March and July 1979.

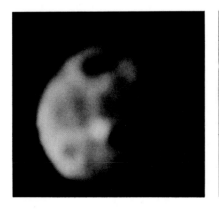 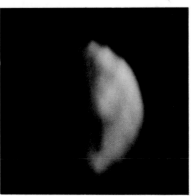 

Not surprisingly, the four larger moons were the first to be discovered, by Galileo, who saw them through his telescopes in 1609. In outwards order from the planet, they are called Io, Europa, Ganymede and Callisto. The fifth member of the inner system, the innermost Jovian moon *Amalthea*, was found by the American astronomer Edward Barnard in 1892. But other than its estimated diameter, 240 kilometres, and its orbital period, less than 12 hours, little is yet known about this closest of Jupiter's lunar companions.

Io, the second Jovian moon, is only a little bigger than ours. It is also more colourful in that it has reddish polar caps – another Barnard discovery. Its close (348,000 kilometres above the clouds) equatorial orbit puts Io in shadow for some part of each of its approximately 42-hour circuits of the giant planet. When it emerges, the brightness of the moon can sometimes appear enhanced, as though clouds or frost had formed while it was hidden from the sun. Whether these effects are due to an atmosphere remains uncertain but in 1974 the American astronomer Robert Brown found Io to be surrounded by a yellow haze due to the presence of sodium atoms. This observation complemented an earlier suggestion that Io, a dense (3·5 times denser than water) and rocky body, is coated with salt flats. Another curious feature of this moon is its not yet understood relationship with Jovian magnetosphere radio-frequency transmissions.

With an estimated diameter of some 3,100 kilometres, *Europa* is a little smaller than our moon. Its brightness suggests a surface layer of gravelly soil and frost. A moderately dense (3·0) body, Europa may have icy polar caps. At an altitude above the Jovian cloud tops of about 600,000 kilometres, it orbits Jupiter in just over 85 hours.

Ganymede, the largest Jovian satellite, is bigger than the planet Mercury. Its estimated diameter, 5,270 kilometres, makes it the third largest moon in the solar system (after Neptune's Triton and Saturn's Titan). Around one million kilometres from Jupiter, it circles the planet in just over a week. Its comparatively low density (about 2) suggests an icy composition.

Callisto, the outermost moon of the inner system, is also a large body. With an estimated diameter of 5,000 kilometres, it orbits Jupiter at an average distance of just under 1,900,000 kilometres from the centre of the planet in less than 17 days. An icy, dust-covered world, it is the least dense (1·65) and darkest of the Galilean satellites.

At about $9\frac{1}{2}$ million kilometres beyond the orbit of Callisto lie the intersecting paths of the four middle-system moons. Unknown until the twentieth century, they were all found by Americans. *Himalia*, the largest (with an estimated diameter of 100 kilometres) was first seen by Charles Perrine in 1904. A year later he found *Elara* (estimated diameter 30 kilometres). In 1938 Seth Nicholson discovered *Lysithea* (estimated diameter 20 kilometres) and in 1974 Charles Kowal found *Leda* (estimated diameter 15 kilometres). All four take about 8 earth months to circle Jupiter.

Spread from the fringes of the middle moon system to a distance of as much as 30,000,000 kilometres from Jupiter, the looser weavings of the orbits of the outer lunar system are another twentieth-century discovery. In 1908 the British astronomer Philibert Melotte found *Pasiphae*. Nicholson discovered the others – *Sinope* in 1914, *Carme* in 1938 and *Ananke* in 1951. Similar in size, they are all thought to have diameters close to 20 kilometres. They take between just under 21 and 25 months to circle Jupiter and unlike most of the other moons in the solar system and all the planets, their orbits are retrograde. It is unlikely, therefore, that they were formed along with Jupiter. They may well be captured asteroids.

In all probability this family tree of Jovian moons will continue to grow. In 1975 Charles Kowal tentatively identified a fourteenth moon.

On the following pages, a sequence of four paintings shows Jupiter as it might appear in the 'sky' of three of its inner-system and one of its middle-system moons.

The first shows the planet looking large over ochre salt flats on Io. In the second, a view from the frosty surface of Europa, both Jupiter and Io can be seen. The third, a view from Callisto, outermost and lightest (least dense) of the four giant moons, shows the distant sun, Jupiter and Ganymede. The last painting shows the planet from the surface of Himalia, the second nearest and largest of the interweaving middle-system moons.

SATURN

First identified in 1659 by the Dutch astronomer Christiaan Huygens, the rings of Saturn are among the more spectacular sights to be seen through a telescope. Spanning 275,000 kilometres, they circumscribe the 120,000-kilometre-diameter planet in its equatorial plane. Radar studies show them to be made up of bands of small solid bodies – composed, perhaps, of rock and ice.

Known as Ring A, the 16,000-kilometre-wide outermost hoop is not as reflective and therefore probably less dense than the middle Ring B which is around 28,000 kilometres wide. These two rings are separated by a 4,000-kilometre gap called Cassini's division, for the Italian-born astronomer Jean-Dominique Cassini who described it in 1675. Its presence was explained in 1867 by the American mathematician James Kirkwood. Like the gaps in the asteroid belt, he reasoned, those in Saturn's ring system are due to the gravitational effects of other orbiting bodies. An object circling Saturn within Cassini's division would have a period equal to half that of one of the planet's inner moons, Mimas, the repeated close approaches of which would tend to deflect that object into another orbit. Other noticeable breaks in the ring system are caused by other moons and fractions of lunar periods.

Within Ring B is the elusive 17,000-kilometre-wide Ring C which rarely shows up well in photographs. Like Ring A, it is much less dense than Ring B. In 1969 a Ring D was tentatively identified. Separated by a substantial gap from the inner edge of Ring C, it is very tenuous. It is also possible that there is an outward extension of the rim of Ring A beyond what can be seen through an earthbound telescope.

Saturn's axis is tilted at an angle of $26\frac{3}{4}°$ with respect to the plane of its orbit around the sun. So, like the earth, it presents alternating hemispheres as it progresses around it. And like that of the earth, Saturn's orbit is an ellipse. The planet's distance from the sun varies between 9 and 10 astronomical units. Also like that of the earth, the southern is the sun-facing hemisphere at perihelion when the distance is least and the planet's orbital speed is greatest. Thus the southern Saturnian 'summer' is shorter than its northern counterpart. And for the same reasons we see the south pole and the rings from below for just $13\frac{3}{4}$ of the $29\frac{1}{2}$ years it takes Saturn to complete one orbit of the sun. For the other $15\frac{3}{4}$ years it is the turn of the north pole and the rings are seen from above. During the Saturnian 'spring' and 'autumnal' equinoxes, the rings appear edge-on and, because they are only a few kilometres thick, they can only be seen with a large telescope. This happened in 1966 and will do so again in 1980 when the northern hemisphere will begin to tilt sunwards.

Like Jupiter, Saturn has a hot, probably rocky core from which it radiates more heat than it receives from the sun. But this does not mean

South is up in this composite of telescopic photographs of Saturn. They were taken when the planet's southern hemisphere and the southern face of its rings were tilted sunwards. In 1980 the rings will appear edge on for a while before gradually revealing their northern face as the northern becomes the sunwards-facing hemisphere.

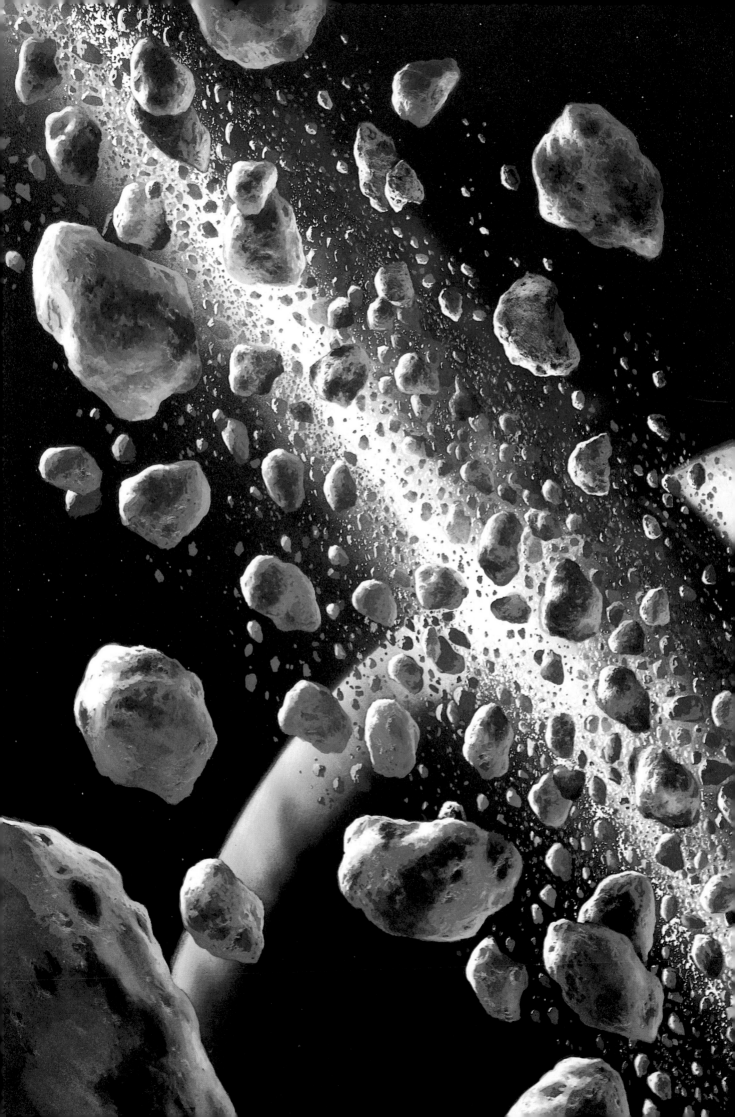

that the whole planet is warm. Saturn's core is thought to be surrounded by an outer coating of ice within a chilling shell of 'metallic' hydrogen. For its size, Saturn is very light. With an average density of 0·7, the planet would float in water. So these three inner layers can only account for 5 or 6 per cent of the total volume, most of which is believed to be a foggy mix of hydrogen, helium, ammonia and methane with cloud top temperatures in the minus 180 °C range.

Another rapidly spinning body – a Saturnian day lasts only 10¼ hours – the planet is moulded by centrifugal forces so that its equatorial diameter is 12,500 kilometres greater than its polar diameter. Like Jupiter's the visible surface of Saturn is banded. A broad pale yellow to whitish equatorial zone is flanked by darker narrower equatorial belts. Other bands are less permanent, although the polar regions usually appear less reflective and bluer than adjacent latitudes. Spots on Saturn are rare, which suggests that Saturnian weather is much less tempestuous than that of the Jovian atmosphere.

The planet has ten known moons whose orbits, with the exception of that of the outermost, Phoebe, are much more orderly than those of the Jovian moons. Their average distances from the centre of Saturn range from less than 160,000 kilometres to nearly 13,000,000 kilometres. None of them is less than 200 kilometres across – although, at a range of never less than 8 astronomical units, smaller bodies are very hard to detect with even the best earthbound telescopes. Titan, the largest, is the second biggest moon in the solar system. Like all moons, whose rotations are known, Saturn's are spin-orbit coupled so that only one hemisphere ever faces the planet.

Janus, the innermost Saturnian moon, is also the 'newest', in that it was found as recently as 1966 by the French astronomer Audouin Dollfus.

Mimas, the second moon in outward order,

The rings of Saturn. Earth-based radar measurements suggest that these are made up of rocky debris and ammonia ice fragments. In this painting the body of the planet is seen along the plane of this orbiting avalanche which may be the result of the disintegration of one or more erstwhile moons.

was found in 1789 by the Hanoverian astronomer Sir William Herschel. It has an estimated diameter of 500 kilometres.

Enceladus, another 1789 Herschel find, is some 600 kilometres in diameter.

Tethys, found in 1684 by Jean-Dominique Cassini, has an estimated diameter of 1,040 kilometres.

Dione, also found by Cassini in 1684, has an estimated diameter of 820 kilometres.

Rhea, found by Cassini in 1672, is some 1,580 kilometres across.

Between 1·0 and 1·5, the densities of the six inner Saturnian moons suggest icy compositions. Also a frosty sphere, the seventh, *Titan* was found by Huygens in 1655. Larger than the planet Mercury, this 5,830-kilometre-diameter moon is remarkable for its possession of an atmosphere which was discovered in 1944 by the American astronomer Gerard Kuiper. Composed of methane and hydrogen, filled with reddish clouds, and around 150 kilometres deep, the gaseous envelope surrounding Titan gives the moon an estimated surface pressure of between a tenth and one earth atmosphere. Leakage of this atmosphere, due to the moon's low gravity, may give rise to a gas ring around the planet. To make good this loss, the atmosphere may be replenished by interior thermal activity. Thus while the surface temperature is normally around minus 150 °C, it is conceivable that periodic local warming creates methane lakes which gradually evaporate into the atmosphere. The biochemical significance of such circumstances is as yet unknown but is clearly of considerable interest.

Hyperion, first seen in 1848 by the American astronomers William Bond and his son George, has an estimated diameter of 500 kilometres.

Iapetus, another Cassini find – he first saw this moon in 1671 – is about 1,600 kilometres in diameter. One side – that which always faces away from the direction in which the moon is travelling in its orbit – is six times brighter than the other. The density of this satellite is estimated to be three times that of water, so the dark side is thought to be exposed rock. The lighter, trailing hemisphere is probably snow-covered.

Phoebe, the outermost moon, found in 1898 by the American astronomer William Pickering, is the smallest known Saturnian satellite with an estimated diameter of 200 kilometres.

Saturn's moons

This diagram gives an oblique view of the orbits of Saturn's ten known moons. (At the time of writing the discovery of an eleventh moon, apparently circling the planet just beyond the rings, awaited confirmation.)

The six inner moons have fairly circular, near-equatorial orbits. Those of Titan, the largest moon, and Hyperion are more elliptical. That of Iapetus is nearly circular but inclined nearly 15° to the plane of the planet's equator. That of Phoebe is elliptical, inclined and retrograde. In the summer of 1979 Phoebe's orbit is due to be crossed by Saturn's first earthly visitor, the American spacecraft

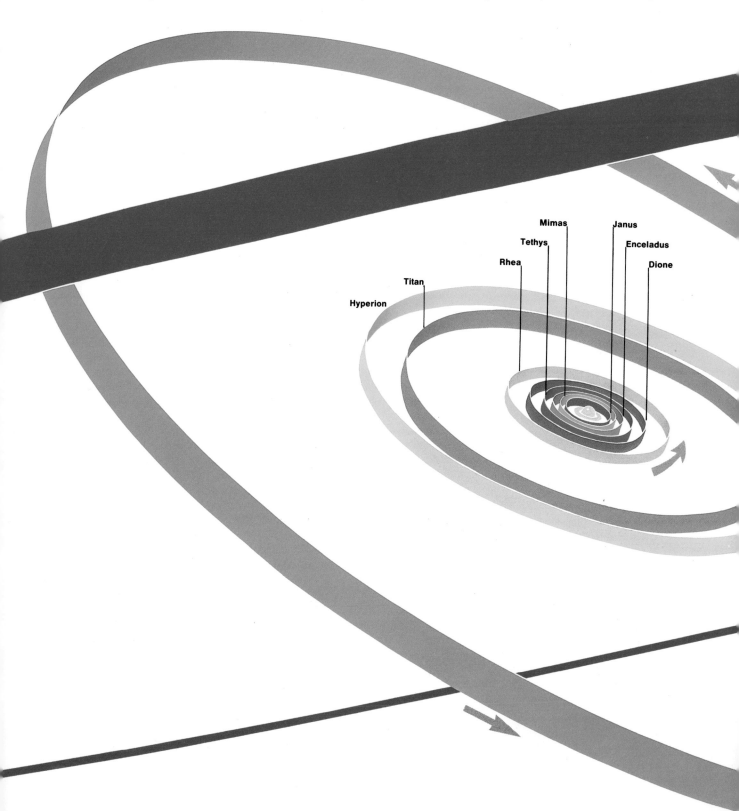

Pioneer 11. The journey, via Jupiter, will have taken six and a half years. Voyagers are also due, in late 1980 and summer 1981.

In the table below, diameters and distances are in kilometres, eccentricity is a measure of how well a moon's orbit fits a circle (○) and orbital periods are relative to the stars.

Moon	Diameter	Average distance from Saturn	Orbital eccentricity	Orbital inclination	Orbital period days	hours	minutes
x Janus	300	159,000	0·000	0° 0'		17	55
i Mimas	500	186,000	0·020	1° 31'		22	37
ii Enceladus	600	238,000	0·005	0° 1'	1	8	53
iii Tethys	1,040	295,000	0·000	1° 6'	1	21	18
iv Dione	820	378,000	0·002	0° 1'	2	17	41
v Rhea	1,580	527,000	0·001	0° 21'	4	12	25
vi Titan	5,830	1,222,000	0·029	0° 20'	15	22	41
vii Hyperion	500	1,483,000	0·104	0° 36'	22	6	38
viii Iapetus	1,600	3,560,000	0·028	14° 43'	79	7	56
ix Phoebe	200	12,952,000	0·163	30°	550 Retrograde		

Roman numerals reflect the order in which the moons were discovered; xi awaits confirmation.

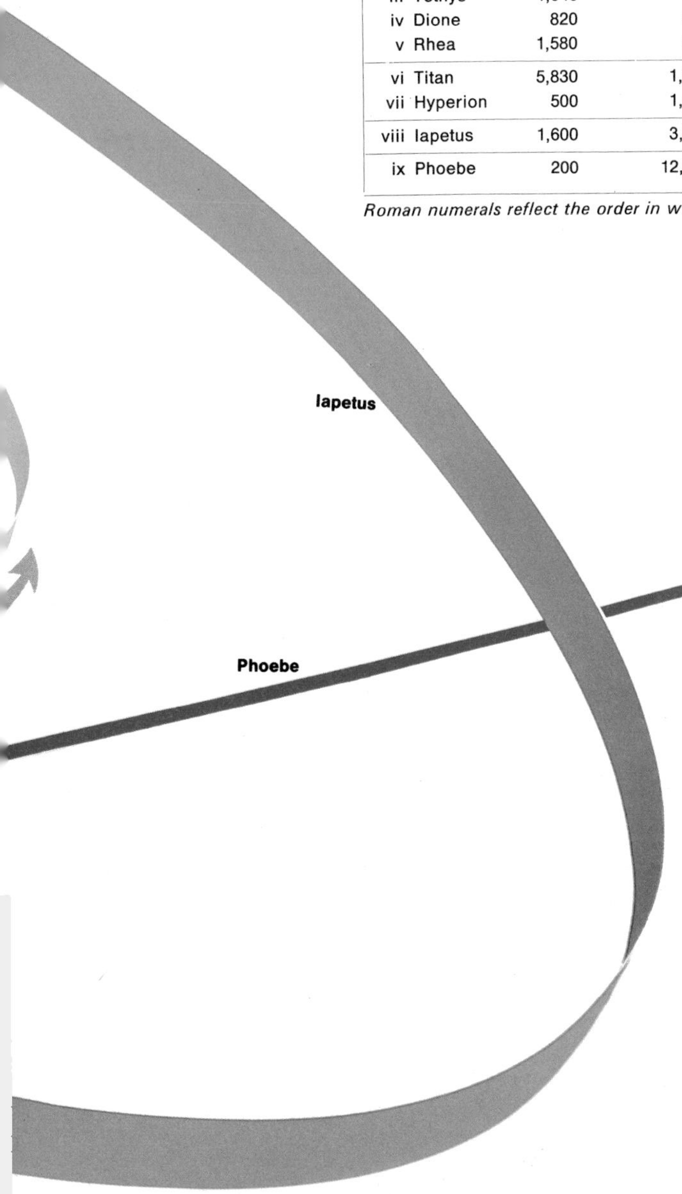

Iapetus

Phoebe

Starting overleaf, a series of six paintings shows Saturn as it might appear from the surfaces of increasingly distant moons. The rings dominate the view from Mimas (1) which can also be seen in the 'sky' of Enceladus (2). Along with the distant sun (left), Enceladus is also visible in the view from Tethys (3). The next painting (4) shows Dione from Rhea. From Titan (5), the moon with an atmosphere, the planet is reflected in a lake of methane. The final view (6) shows Saturn from Iapetus.

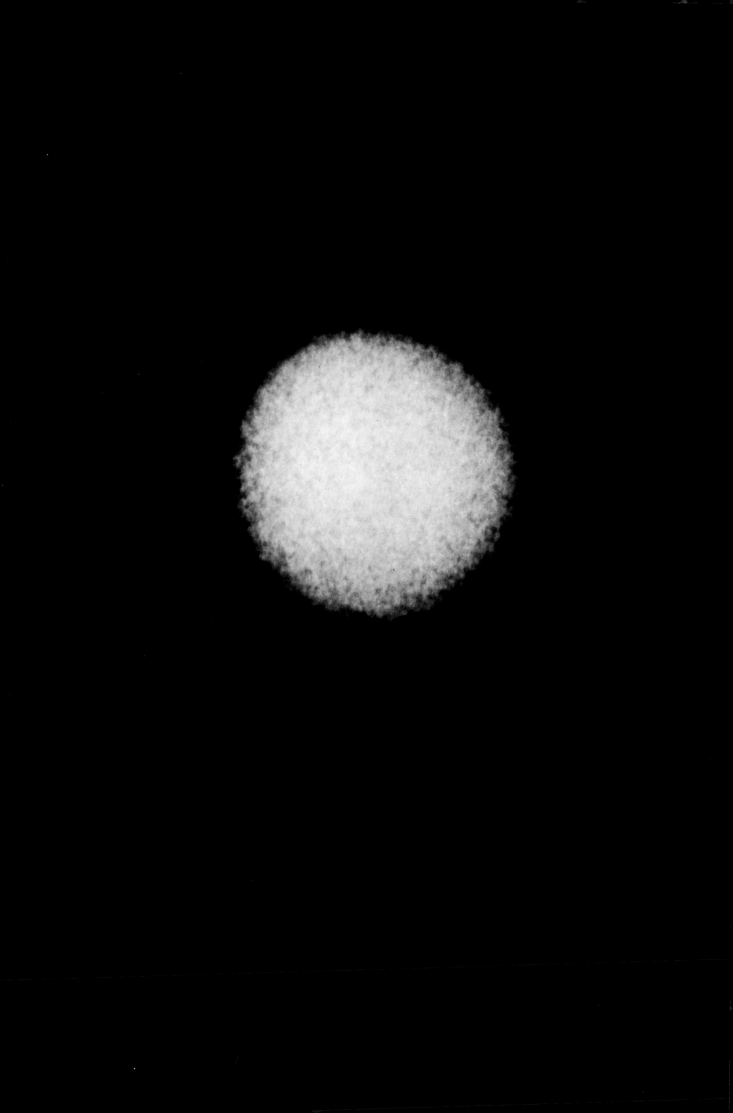

URANUS

Half the size of Saturn and twice as far from the sun, Uranus can be seen, albeit only just, with the naked eye. That it was still undiscovered 170 years after the invention of the telescope might seem surprising. It was certainly seen in 1690 but not recognized for what it is – a wanderer among millions of fixed stars – until 1781. In that year Sir William Herschel recorded 'a curious either nebulous star or perhaps a comet'. For some time called the 'Georgian Planet' for Herschel's patron King George III, Uranus was finally named for the father of Saturn in the Roman pantheon. It is of more than passing interest, two centuries later, that Herschel thought, for a while, that he could see rings around the planet.

Denser (1·2) than Saturn, Uranus is believed to possess a hot rocky core, not so very different in size to that of the ringed planet, surrounded by an icy mantle. Its chilly (minus 210 °C) outer layers are thought to be composed of a mixture of hydrogen, helium, ammonia and methane. The last absorbs sunlight preferentially from the red end of the spectrum, giving the planet its greenish tinge.

The remotest known planet until 1846, Uranus is never less than 2,500 million kilometres – over 17 astronomical units – from the earth. Enigmatic in this composite telescopic view, Uranus – like Saturn – has rings. They were discovered in March 1977 when they interrupted light from a star as the planet passed in front of it. Photographs of the rings may be returned by an American Voyager craft in 1986.

Like that of Venus, the rotation of Uranus is retrograde so that, seen from a point above its north pole, the planet appears to spin clockwise. Like Jupiter and Saturn, Uranus is flattened so that its equatorial diameter is 3,000 kilometres greater than its polar diameter. Unlike that of any other planet or moon, the Uranian axis is tilted an exceptional 82° so that it lies very nearly in the plane of the planet's orbit. Viewed from above this plane, the planet therefore appears to spin backwards as it swings around the sun.

Not surprisingly, the Uranian calendar is correspondingly bizarre. The planet's exact rotation period is as yet uncertain but thought, with reference to the stars, to be between 17 and 23 hours. But if a day is taken to be the interval between one sunrise and the next, variations with latitude and season will be enormous. At the Uranian north pole, for example, such a 'day' will last 84 earth years, the time it takes the planet to complete a circuit of the sun.

In March 1977 Uranus was found to have a system of at least five narrow rings lying in its equatorial plane at between about 18,000 and 25,000 kilometres from the planet's surface. They were discovered by a team of astronomers whose equipment included an American airborne observatory flying across the southern Indian Ocean south-west of Australia. The team were set to observe the passage of the planet in front of a star. The astronomers were hoping to learn something of the nature of the Uranian atmosphere as the star set and rose behind the passing planet. And by comparing timings from several sites, they hoped to get a better estimate of the planet's diameter and

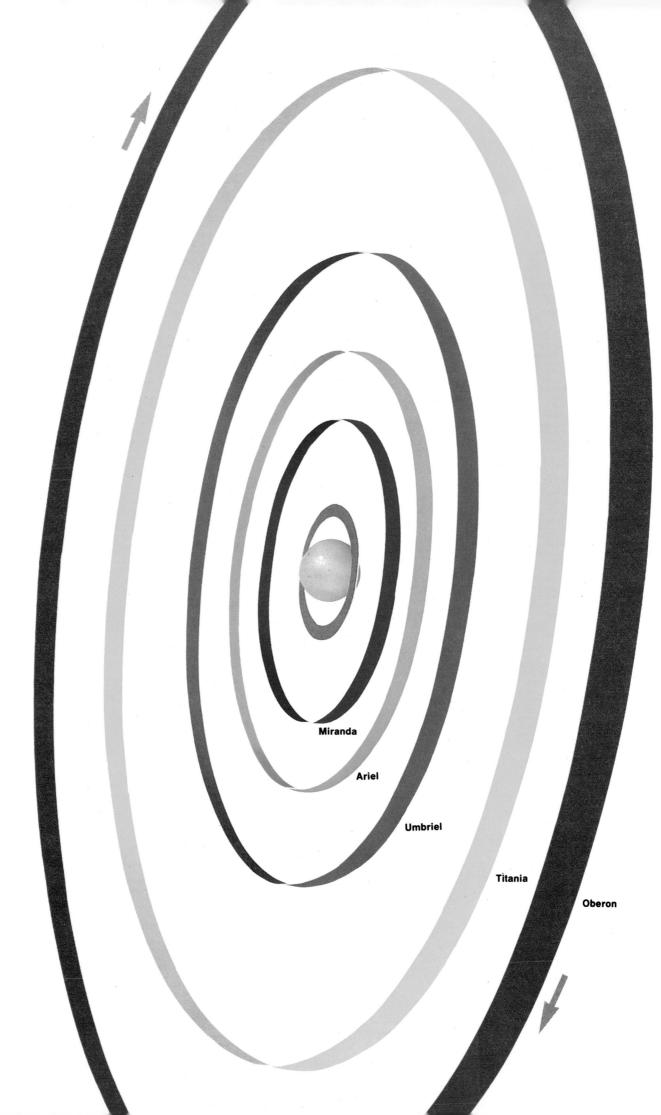

Miranda

Ariel

Umbriel

Titania

Oberon

oblateness. The first indication of the presence of rings came when Uranus was about to pass in front of the star which, to the surprise of the watching astronomers, faded and brightened on five separate occasions a few minutes before the planet obscured it. Repetition of the fading and brightening sequence in reverse order, after the star had emerged from behind Uranus, confirmed the discovery.

The rings are much narrower than those of Saturn and will probably be impossible to photograph until the early 1980s when the Americans are due to launch a 2·4-metre telescope – the Space Telescope – into earth orbit. A closer look may be obtained some time in 1986 – from a Voyager spacecraft due to fly past the planet via Jupiter and Saturn.

Also in the plane of the planet's equator is the Uranian family of five known moons. They all have near-circular orbits and move in the same direction as the planet's spin which suggests that they were formed at the same time as their parent. The outermost two, Oberon and Titania, were found by Herschel in 1787. *Oberon* has an estimated diameter of 1,600 kilometres. *Titania* is about 1,800 kilometres across. The next two moons were discovered by the English astronomer William Lassell in 1851. *Umbriel* has an estimated diameter of 1,000 kilometres. *Ariel* is half as big again. The innermost moon, *Miranda*, was first seen in 1948 by the American astronomer Gerard Kuiper. Also the smallest, it has an estimated diameter of 550 kilometres.

Mathematical analysis of the arrangement of the rings in relation to the motions of the moons has led to the suggestion that there might be another moon lying in an equatorial orbit within that of Miranda. Calculations indicate that the postulated sixth moon should be about the size of Miranda and that its average distance from Uranus should be around 100,000 kilometres.

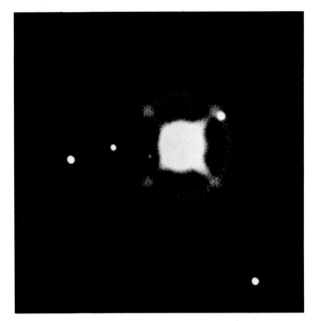

The photograph above shows Uranus and its moons. The apparent ring is an optical effect – a consequence of the long exposure needed to obtain images of the moons. A real ring system was discovered in March 1977. Its position and dimensions are indicated in the diagram on the left which shows Uranus from a point in its orbit which it has just passed (the sun is to the left). Beyond the rings (orange band) are the orbits of the five known moons.

The axis of the planet is inclined some 82° to the vertical and its direction of spin, when viewed from a point above its north pole, is clockwise and retrograde. The satellites' equatorial orbits are also retrograde in the sense that the moons travel in the same direction as the planet's spin. The motion of the whole system around the sun involves, therefore, various degrees of back spin.

Overleaf, the first of two paintings shows Uranus viewed along the plane of its newly discovered rings. The sun appears behind Miranda, the innermost of five known moons.

The second illustration shows the planet rising above the horizon of Miranda. The sun (top right) appears in the background.

Moon	Diameter	Average distance from Uranus	Orbital eccentricity	Orbital inclination		Orbital period		
						days	hours	minutes
v Miranda	550	131,000	0·017	0°	0′	1	9	55
i Ariel	1,500	192,000	0·003	0°	0′	2	12	29
ii Umbriel	1,000	267,000	0·004	0°	0′	4	3	28
iii Titania	1,800	438,000	0·002	0°	0′	8	16	56
iv Oberon	1,600	586,000	0·001	0°	0′	13	11	7

Roman numerals reflect the order in which the moons were discovered.

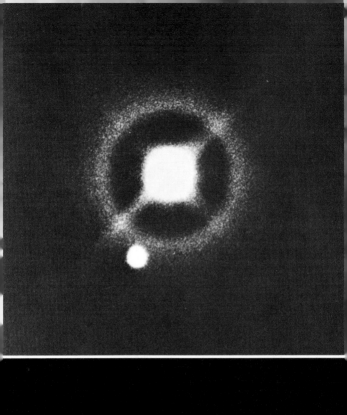

A telescopic view of Neptune and its inner satellite, Triton. The apparent ring around the planet is an optical effect. With a diameter of about 6,000 kilometres, Triton is possibly the solar system's largest moon.

NEPTUNE

In 1841 the English astronomer John Adams, whose working life was mostly devoted to a study of the dynamics of the solar system, became interested in the fact that there were important discrepancies between the calculated and observed motion of Uranus. By 1845 his computations not only indicated that the seventh planet was being disturbed by the gravitational attractions of an unseen eighth, they also suggested where, in the night sky, this new planet might be found. As always the availability of good telescopes was limited, so it was up to the then Astronomer Royal, Sir George Airy, to organize a search. But he did not. In France meanwhile, Urbain Le Verrier had independently reached much the same conclusions. Learning of this, Airy took belated action but although the Cambridge astronomer James Challis saw and recorded the planet on 4 August 1846, he failed to compare his observation notes with those of the previous night and did not therefore recognize it for what it was. Using Le Verrier's calculations, Neptune was found by two Berlin astronomers Johann Galle and Heinrich d'Arrest.

As well as casting a shadow across the reputations of Airy and Challis, the German success marked the end of Johann Bode's winning streak. Against a predicted 38·8 astronomical units, Neptune was found to lie in a near-circular orbit at a mean distance from the sun of just over 30·0 astronomical units. Although its density (1·7) is greater, Neptune is much the same size as Uranus. Its estimated diameter is 49,500 kilometres. Its internal structure is also probably similar to that of its inside neighbour – a warm rocky core encased in ice with a thick external blanket of hydrogen, helium, ammonia and methane, a mixture which gives the visible surface its bluish-green colouring. Cloud-top temperatures are thought to be minus 220 °C.

The tilt of Neptune's axis is just less than 29° so the length of the solar day is not unlike the rotation period of the planet, which is believed to be between 19 and 23 hours. It takes Neptune just under 165 earth years to complete one of its own so there are at least 68,000 Neptunian days in a Neptunian year.

The nearest moon is *Triton*. It was found by the English astronomer William Lassell within a month of the discovery of the planet. Bigger than the planet Mercury, it has an estimated diameter of 6,000 kilometres and is the largest moon in the solar system. Its orbit around Neptune is circular, retrograde and inclined 20° to the planet's equatorial plane. At a distance always within a few per cent of 355,000 kilometres from the centre of Neptune, Triton takes 141 hours to circle the planet.

Neptune's second lunar companion is also exceptional. Discovered in 1949 by the American astronomer Gerard Kuiper, *Nereid* has a highly eccentric orbit so that its distance from Neptune can be as little as 1,600,000 kilometres or as much as 9,600,000 kilometres. It has an estimated diameter of 500 kilometres and takes 360 earth days to circle the planet.

The painting overleaf shows Neptune as it might appear from Triton's ice-bound surface. Neptune may be reached by an American Voyager spacecraft sometime in 1989 or 1990, by way of Jupiter, Saturn and Uranus.

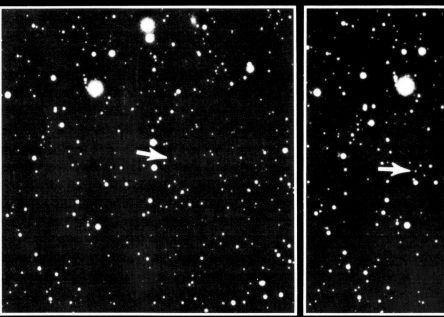

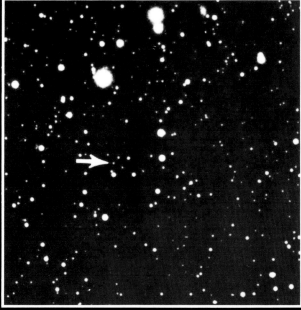

These two images show the same star field photographed on 21 and 29 January 1930. When he compared them in February that year, the American astronomer Clyde Tombaugh found Pluto (arrowed), which had moved eight days along its orbit against the stellar background.

The painting overleaf shows the sun as it would appear in the 'sky' of Pluto. While Pluto is sometimes more than seven billion kilometres from the sun, it will cease to be, in 1979, the remotest planet – for twenty years it will lie inside the orbit of Neptune.

PLUTO

With the finding of Neptune, the successors of Adams and Le Verrier set about their computations with renewed confidence. But the planets still refused to run according to their timetable. The eighth, it was soon apparent, had an unseen accomplice. Precise measurements of the motions of the known outer planets indicated yet another gravitational presence. So once again predictions were made and telescopes pointed at the fringes of the solar system. In 1905 the cartographer of the Martian canals, Percival Lowell, began his quest. His *Memoir on a Trans-Neptunian Planet* was published in 1915. Although he died the following year, it was with a purpose-built telescope at his observatory that the American astronomer Clyde Tombaugh found Pluto in 1930. The ninth planet was named as much for the ancient Greek ruler of the dark underworld as for the first two letters of his name which are also Lowell's initials.

Like the solar system's remotest known moon, the farthest known planet flies a highly eccentric orbit, the plane of which – unlike those of the other planets – is considerably tilted (17.2°) to that of the earth's orbit around the sun. Pluto's greatest distance from the sun at aphelion is 49.3 astronomical units or 7,375,000,000 kilometres but its average distance is only 39.4 astronomical units. Its least distance at perihelion is a mere 29.6 astronomical units which brings the planet inside the orbit of Neptune for twenty years of its 248-year journey around the sun. Such an inwards crossing is due in January 1979. There is no chance of a collision however, for quite apart from the rarity of occasions when the two planets are at the coincident points along their orbits, the current spatial relationship between them is that of flyover and underpass. According to some astronomers this may not always have been the case. The disorderly state of the orbits of Neptune's moons, they say, suggests a dramatic past – a falling apart of an earlier lunar family with the subsequent departure of Pluto from its original Neptunian orbit.

Pluto is so far from the earth that, in the eye-pieces of even the most powerful telescopes, it appears, not as a reflective disc as do other planets, but as an enigmatic gleam, like the reflection of a bright light by a billiard ball. By studying cyclic changes in its brightness, astronomers have been able to deduce an indicated rotation period of 153 hours but this value is uncertain.

Pluto's diameter is thought not to exceed 6,000 kilometres. But it may turn out to be as little as half that figure. In July 1978 an American astronomer, James Christy, pointed out that in several recent photographs of Pluto it appears to have a somewhat elongated shape. This, he argued, could be accounted for if the image is seen, not as that of a single body, but as that of a 3,000-kilometre-diameter planet with a 1,200-kilometre-diameter moon orbiting the planet at a distance of 17,000 kilometres. Pluto, it was suggested, is little more than a snowball of frozen gases with a relatively large moon, dubbed Charon by Christy, of similar composition.

Both bodies may have been liberated from Neptune by Planet X – the 'missing' tenth planet which may have passed through Neptune's satellite system long ago on its way to its present hiding place, 50 to 100 AU from the sun.

COMETS

Above, beyond and below the orderly plane of interplanetary space lie the looping runs of the wilder members of the sun's family – hard-centred clouds of ice and dust, showers of rocks – the comets and the meteoroids. Many of the latter are grouped in rings like those of Saturn and Uranus but of much greater circumference and eccentricity, and discontinuous like a ragged string of beads. A continuous segment of such a ring – where the rocky fragments are relatively close together – is often called a meteor stream although *meteoroid* is the preferred term for ring components while they are still in space. Where their orbit is intersected by that of the earth, they may enter the latter's atmosphere; their partial consumption by friction burning can then produce visible *meteors*. Surviving solid fragments are called *meteorites*.

Meteoroids large enough to pass through a meteor stage add around 2,000 tonnes to the earth's mass every year but the amount of dust entering the earth's atmosphere annually, in the form of micrometeorites, is estimated to be as much as 200,000 tonnes. Very big meteorites produce fireballs visible in daylight, and often recoverable fragments.

Whereas meteors and fireballs are incandescent, the cold fire of comets is not due to friction but to sunlight reflection and gas fluorescence. Often associated with meteoroid rings, comets have solid nuclei, composed of

frozen carbon monoxide, methane and water mixed with dust. Thawing, as the comet advances towards the sun, produces an enormous tenuous envelope of ice crystals, dust particles and ionized gas which may grow to as much as 1,800,000 kilometres in diameter. Blown by the solar wind, a stream of this luminous vapour forms a tail which may be tremendously long – that of the 'great comet' of 1843 spanned more than 2·1 astronomical units, some 320,000,000 kilometres. The tail always flows away from the sun so that, after perihelion, a retreating comet travels tail-first.

Recorded since 87 BC or earlier, Halley's, perhaps the most famous of the comets, is not however typical. Its highly elliptical seventy-six-year orbit encompasses those of all the planets except Mercury. Its next close turn around the sun – at a range of 0·6 astronomical units – is due in February 1986. But interplanetary comets form a tiny minority of the total population.

The real home of the comets – estimated by some astronomers to number 100 billion – lies half-way to the nearest stars, in the remote gravitational fringes of the solar system, between 20,000 and 60,000 astronomical units from the rim of its planetary core. From where the sun, hardly brighter than other stars, would seem to rock from side to side as do the other suns which play pivot to unseen satellites. And from where the attention of a sentient being with a knowledge of radar and radio communication might be drawn by one of nine planets, a noisy little world shedding spores into interstellar space – capsules bearing emblems of earth.

The comet Bennett photographed in April 1970. Halley's comet, recorded since 87 BC or earlier, visits the inner solar system every seventy-six years. The painting overleaf shows it in the Mercurian night 'sky' in 1986.

219

ACKNOWLEDGEMENTS

For their considerable and generous assistance, the author is deeply grateful to many wise and helpful people.

For reading and commenting on whole or part of the text: Professor Carl Sagan, Richard French and David Pieri of the Laboratory for Planetary Studies, Cornell University; Don Bane of the Jet Propulsion Laboratory (JPL), California Institute of Technology; Peter Gill of the Royal Astronomical Society and John Murray of the University of London Observatory. It is my fault, not theirs, if any gross error remains and it should be noted that in some places I have favoured one interpretation of the known facts over others.

For finding and supplying photographs and other data with which diagrams have been prepared: the Office of Public Affairs of the US National Aeronautics and Space Administration (NASA) – especially Les Gaver and Margaret Ware of NASA Headquarters; Peter Waller, Rosemarie Partolan and Debora Silberberg of Ames Research Center; Ed Mason, Joseph McRoberts and Don Witten of Goddard Space Flight Center; Robert J. MacMillin and Bonnie Cantrell of JPL.

I am also indebted to Sadie Alford of the Novosti Press Agency, Peter Baylis of the University of Dundee, Phil Bolger of the Office of the US Secretary of Transportation, Fred Durant of the Smithsonian Institution, Evald Karlsen of Victor Hasselblad Aktiebolag, Enid Lake of the Royal Astronomical Society, Stephen Larsen of the University of Arizona, Charles Pike of US Department of Defense, Nigel Press of Nigel Press Associates, Jeannette Seaver of Richard Seaver Books, Andrew Sinclair of Her Majesty's Nautical Almanac Office and Gordon Vaeth of the US National Oceanic and Atmospheric Administration. A complete list of picture sources appears below.

For their professional time and encouragement: Gordon Andreasen, Peter Dilonardo and Fred Lavery of the US Geological Survey, my fellow science writer, Ian Ridpath and Sir Patrick Skipwith of the Bureau de Recherches Géologiques et Minières.

For their skilful and imaginative conversion of rough sketches into magnificent diagrams: Alastair Campbell, Edward Kinsey and the other members of the QED studio.

For his advice and her preparation of the manuscript: my editors Peter Carson and Esther Sidwell. For his design of the book: Fred Price. And for his care with its production: Alan Smith.

To all these people and to my partner, Ludek Pesek, thank you.

ILLUSTRATIONS

In the list of illustration sources below, where catalogue references follow page numbers, the following abbreviations have been used: ERTS (Earth Resources Technology Satellite – now called Landsat), JPL (Jet Propulsion Laboratory), NASA (National Aeronautics and Space Administration) and RAS (Royal Astronomical Society).

12	NASA 72-H-946 (Hale Observatories)
24	top NASA 62-OSO-4; bottom NASA 62-OSO-3
25	RAS 547 (Royal Observatory, Greenwich); inset left RAS 637 (Royal Greenwich Observatory, Herstmonceux); inset right RAS 636 (Royal Greenwich Observatory, Herstmonceux)
28	NASA 70-H-367
29	NASA 70-H-460
30-31	NASA 72-H-1053
30	NASA 70-H-462
31	NASA 65-H-1965
32-3	NASA 74-HC-543
33	NASA 63-Ionosphere-6
34-5	NASA 62-OSO-1 (Mount Wilson – Palomar Observatories)
35	inset NASA (courtesy Victor Hasselblad Aktiebolag)
36	NASA 65-H-2012
37	top Hale Observatories SP 65b; bottom NASA 74-HC-260
38-9	NASA 74-HC-179
38	inset NASA 74-H-40
40-41	NASA 74-HC-30
40	inset left NASA 74-H-39; inset right Hale Observatories SP 58
42	NASA 74-HC-33
43	top NASA 73-HC-752; bottom NASA 73-HC-751
44-5	NASA 73-HC-626
48	NASA 71-H-955
49	top NASA 73-HC-908; bottom NASA 73-HC-907
50	United States Air Force
52	NASA 74-H-239 (JPL P-14470)
54-5	sequence left to right RAS 718-20
54	top RAS 448
55	NASA 70-H-1657 (Lunar and Planetary Laboratory, Arizona)
59	NASA 74-H-239 (JPL P-14470)
60	NASA JPL P-15119
61	top NASA 74-H-406; bottom NASA 74-H-240 (JPL P-14472)
64-5	NASA JPL P-14468
66	NASA JPL P-14580
68	top NASA JPL P-15427; bottom NASA JPL P-14473
69	NASA 74-H-536
70	NASA JPL P-15046
71	NASA 75-H-1085 (JPL neg. 321-684)
72-3	NASA JPL P-15197
76	NASA JPL P-14400
78	NASA
82	NASA JPL P-14422
83	top NASA 76-H-693; bottom NASA 76-H-694
86-7	Novosti Press Agency
90	earth NASA (courtesy Victor Hasselblad Aktiebolag); moon NASA 72-HC-442
101	top NASA ERTS E 1636-14584 April 1974; bottom E. S. Barghoorn, Harvard University
103	NASA 72-HC-660
104	top United States National Oceanic and Atmospheric Administration, NOAA-2 26-7 August 1974; bottom NOAA-3 24 August 1974
108-9	left NASA ERTS E 1172-17135 January 1973; right NASA ERTS E 1352-17134 July 1973
110-11	NASA ERTS E 1090-18012 October 1972
112-13	top NASA ERTS E 1256-21391 April 1973; right NASA ERTS E 1039-07595 August 1972; bottom NASA ERTS E 1183-06194 January 1973
114-15	top United States National Oceanic and Atmospheric Administration and University of Dundee,
	NOAA-5 18 August 1976; right and bottom NASA ERTS E 2188-10165 July 1975
116-17	top NASA ERTS E 1243-10132 March 1973; right NASA ERTS E 1041-09471 September 1972; bottom NASA ERTS E 1075-09375 October 1972
118-19	left NASA ERTS E 1010-08375 August 1972; right NASA, Landsat 1
120	Peter Ryan
123	top left NASA AS16-116-18599; top right NASA (courtesy Victor Hasselblad Aktiebolag); centre NASA, Lunar Orbiter 4 frame H 127; bottom NASA 67-H-1405
124	top left NASA 67-H-1206; top right NASA 67-H-1406; bottom NASA 66-H-1470
125	top NASA, Lunar Orbiter 5 frame M 105; bottom NASA AS 15-84-11287
126	Lick Observatory L9
129	NASA 72-H-848
130	top NASA 75-HC-159; bottom NASA 72-HC-409
131	top NASA AS 16-116-18690; bottom NASA AS 12-48-7134
132	NASA JPL P-16870
135	top NASA JPL P-17022; bottom NASA JPL P-18598
136	top NASA JPL P-17444; bottom left NASA JPL, Mariner 9; bottom right NASA JPL P-17442
138	centre Lowell Observatory; bottom left NASA JPL P-11110
141	centre Lowell Observatory; bottom right NASA JPL P-17442
142-3	NASA JPL neg. 211-5160
143	NASA JPL P-18078
146	top NASA JPL P-16952; bottom NASA JPL 211-4987
147	top NASA JPL P-17138; bottom NASA JPL P-17698
148-9	top NASA JPL P-17428; bottom NASA JPL P-17430
148	centre left NASA JPL P-18096
149	centre right NASA JPL P-17164
150-51	NASA JPL P-17689; bottom left P-18296; bottom right P-19541
152	NASA JPL P-18086
155	top NASA JPL P-18459; bottom NASA JPL P-17679
156	top NASA JPL P-17599; bottom NASA JPL P-19600
159	top left NASA JPL P-19599; top right NASA JPL P-17873; bottom left P-18162; bottom right P-19133
162	NASA Ames A-73-9175
164	top NASA Ames A-73-9205; bottom NASA Ames A-74-9240
165	top NASA Ames A-73-9188; bottom NASA Ames AC-74-9301
166	NASA Ames AC-74-9206
167	NASA Ames AC-9213
170	NASA Ames A-73-9175
174	top NASA Ames A-73-9279; bottom left to right NASA Ames AC-74-9335, AC-74-9334 and AC-74-9336
184	Stephen Larson, Lunar and Planetary Laboratory, Arizona
202	NASA 70-H-784
203	RAS 573 (Gerard Kuiper, McDonald Observatory)
210	RAS 575 (Gerard Kuiper, McDonald Observatory)
214	Lowell Observatory
218	NASA 70-H-728

INDEX